THE LIBRARY
ST. MARY'S COLLEGE OF MARYLAND
ST. MARY'S CITY, MARYLAND 20686

SCIENTIFIC ALTERNATIVES
TO ANIMAL EXPERIMENTS

ELLIS HORWOOD SERIES IN BIOCHEMISTRY AND BIOTECHNOLOGY

Series Editor: Dr ALAN WISEMAN, Senior Lecturer in the Division of Biochemistry, University of Surrey, Guildford

Aitken, J.	Handbook of Enzyme Active Site Identification*
Ambrose, E.J.	The Nature and Origin of the Biological World
Austin, B. & Brown, C.M.	Microbial Biotechnology: Freshwater and Marine Environments*
Berkeley, R.C.W., *et al.*	Microbial Adhesion to Surfaces
Bertoncello, I.	Human Cell Cultures for Screening Anti-Cancer Rays*
Blackburn, F. & Knapp, J.S.	Agricultural Microbiology*
Bowen, W.R.	Membrane Separation Processes*
Bubel, A. & Fitzsimons, C.	Microstructure and Function of Cells
Clarke, C.R. & Moos, W. H.	Drug Discovery Technologies
Cooke, N.	Potassium Channels
Corkill, J.A.	Clinical Biochemistry: The Analysis of Biologically Important Compounds and Drugs*
Crabbe, M.J.C.	Development of Enzymes for Biotechnology through Protein Engineering
Crabbe, M.J.C.	Kinetics of Enzymes*
Denyer, S. & Baird, R.	Handbook of Microbiological Control*
Dolly, J.O.	Neurotoxins in Neurochemistry
Espinal, J.	Understanding Insulin Action: Principles and Molecular Mechanisms
Eyzaguirre, J.	Chemical Modification of Enzymes
Eyzaguirre, J.	Human Biochemistry and Molecular Biology
Ferencik, M.	Immunochemistry*
Fish, N.M.	Computer Applications in Fermentation Technology*
Francis, J.L.	Haemostasis and Cancer*
Gacesa, P. & Russell, N.J.	Pseudomonas Infection and Alginates: Structures, Properties and Functions in Pathogenesis*
Gemeiner, P. *et al.*	Enzyme Engineering*
Harding, J.	Biochemistry and Pharmacology of Cataract Research: Drug Therapy Approaches*
Horobin, R.W.	Understanding Histochemistry: Selection, Evaluation and Design of Biological Stains
Hudson, M.J. & Pyle, P.L.	Separations for Biotechnology, Vol 2*
Hughes, J.	The Neuropeptide CCK
Jordan, T.W.	Fungal Toxins and Chemistry
Junter, G.A.	Electrochemical Detection Techniques in the Applied Biosciences: Vol. 1: Analysis and Clinical Applications Vol. 2: Fermentation and Bioprocess Control, Hygiene and Environmental Sciences
Kennedy, J.F. & White, C.A.	Bioactive Carbohydrates
Krcmery, V.	Antibiotic and Chemotherapeutic Compounds*
Krstulovic, A. M.	Chiral Separations by HPLC
Lembeck, F.	Scientific Alternatives to Animal Experiments
Palmer, T.	Understanding Enzymes, 2nd Edition
Paterson, R.	Biological Applications of Membranes
Reizer, J. & Peterkofsky, A.	Sugar Transport and Metabolism in Gram-positive Bacteria
Russell, N. J.	Microbes and Temperature
Scragg, A.H.	Biotechnology for Engineers: Biological Systems in Technological Processes
Sikyta, B.	Methods in Industrial Microbiology
Sluyser, M.	Molecular Biology of Cancer Genes
Sluyser, M. & Voûte, P.A.	Molecular Biology and Genetics of Childhood Cancers: Approaches to Neuroblastoma
Verrall, M.S.	Discovery and Isolation of Microbial Products
Verrall, M.S. & Hudson, M.J.	Separations for Biotechnology
Walum, E, Stenberg, K. & Jenssen, D.	Understanding Cell Toxicology: Principles and Practice
Webb, C. & Mavituna, F.	Plant and Animal Cells: Process Possibilities
Winkler, M.	Biochemical Process Engineering*
Wiseman, A.	Handbook of Enzyme Biotechnology, 2nd Edition
Wiseman, A.	Topics in Enzyme and Fermentation Biotechnology Vols. 1, 2, 4, 6, 8
Wiseman, A.	Enzyme Induction, Mutagen Activation and Carcinogen Testing in Yeast
Wiseman, A.	Genetically Engineered Proteins and Enzymes from Yeasts

* *In preparation*

SCIENTIFIC ALTERNATIVES TO ANIMAL EXPERIMENTS

Editor
FRED LEMBECK M.D.
Professor of Pharmacology
University of Graz, Austria

Translator
JACQUI WELCH
University of Southampton

Translation Editor
Dr JOHN FRANCIS
University Department of Haematology, University of Southampton

ELLIS HORWOOD LIMITED
Publishers · Chichester

First published in 1989
ELLIS HORWOOD LIMITED
Market Cross House, Cooper Street,
Chichester, West Sussex, PO19 1EB, England
The publisher's colophon is reproduced from James Gillison's drawing of the ancient Market Cross, Chichester.

Distributors:
Australia and New Zealand:
JACARANDA WILEY LIMITED
GPO Box 859, Brisbane, Queensland 4001, Australia

Europe and Africa:
JOHN WILEY & SONS LIMITED
Baffins Lane, Chichester, West Sussex, England

USA and Canada:
CHAPMAN AND HALL, INC.
29 West 35th Street
New York, NY 10001–2291

South-East Asia
JOHN WILEY & SONS (SEA) PTE LIMITED
37 Jalan Pemimpin # 05–04
Block B, Union Industrial Building, Singapore 2057

Indian Subcontinent
WILEY EASTERN LIMITED
4835/24 Ansari Road
Daryaganj, New Delhi 110002, India

© 1989 F. Lembeck/Ellis Horwood Limited

British Library Cataloguing in Publication Data
Scientific alternatives to animal experiments.
1. Medicine. Laboratory technqiues
I. Lembeck, Fred
610'.28

Library of Congress CIP data avialable

ISBN 0-7458-0589-2 (Ellis Horwood Limited)
ISBN 0-412-02771-2 (Chapman and Hall, Inc.)

Printed in Great Britain by The Camelot Press, Southampton

COPYRIGHT NOTICE
All Rights Reserved. No part of this publication may be reproduced, stored in a retrieval system, or transmitted, in any form or by any means, electronic, mechanical, photocopying, recording or otherwise, without the permission of Ellis Horwood Limited, Market Cross House, Cooper Street, Chichester, West Sussex, England.

Table of contents

Introduction 1
F. Lembeck

1. THE BASIS OF ANIMAL EXPERIMENTATION 4

From discovery to research
From observations in man to research on animals - an historical overview 4
F. Lembeck

From research to cure
The scientific basis of medicine 9
F. Lembeck
        Neurochemistry and Parkinson's disease 9
        Playing with histamine 10
        Substitute for morphia 11
        Clover poison - prophylaxis for thrombosis 11

Sick animals as experimental tools?
Experimental diseases in the animal as a model for therapeutic research 12
F. Lembeck
        Vitamin research 13
        Hormones 14
        Chemotherapy 15
        Genetic defects in animals 16
        Allergies 16

2. DRUG DEVELOPMENT 17

The clinical outlook
Discovery of drugs through clinical observations 17
F. Lembeck
    The beginning of anaesthetics 18
    Heart drugs 19
    Reduction in blood sugar 19
    Treatment of hypertension 20
    Contergan 20

The unexpected result
Findings in animal research: the door to the understanding of function, diagnosis and therapy 21
U. Holzer

Strategic developments
Targeted search for new drugs in animal experiments 25
R. Kilches

From mouse to man
Transferability of the results of animal experiments to man 30
W. Kobinger
    Animal experiments are not transferable 30
    From chance to progress 30
    Transferability: Good - bad 31
    Healthy animals - sick people 31
    Is basic research unprofitable? 32
    The test pyramid 33
    Alternative methods: a reorientation 34
    Cost-benefit ratio - economic aspects 34

3. TOXICOLOGY: SAFETY FOR MAN AND ANIMAL 38

Dependence on animal research
(Drug safety) 39
R. Czok
    The therapeutic effect 40
    Safety 41

Toxicity of lipsticks?
Safety of skin care products, food additives and agents in daily use 44
F. Lembeck

Poisons by the ton!
Safety of industrial chemicals 47
D. Henschler
        Safety limits as a protection principle 47
        New exigencies in work protection: the pressure of innovation 48
        New type of action: genotoxicity 51
        Maximal Workplace Concentrations: for and against 51
        Safety limits for carcinogenic agents? 52
        The dilemma: lack of quantitation? 52

Safety measures
New methods in toxicology: cell culture 54
H. Marquardt
        Aims of toxicology 54
        "Alternative methods" 55
        Acute toxicity 56
        Special toxicities 57

4. METHODOLOGY DEVELOPMENTS 61

From dog to egg?
Large and small laboratory animals 61
G. Skofitsch
        Our relationship with animals 61
        Pure research - applied research 63
        More highly organised animals in applied research 64
        Lower animals in applied research 65
        Safety investigations 65

Experiments on the organs of dead animals
The isolated organ 68
P. Holzer
        The various levels of biomedical research 68
        A look at the past 69
        Potentials and limitations of the methodology of isolated organs 70

Offal for research?
Organs from the slaughterhouse 72
E. Beubler

Human surgical and post-mortem material
Ethically justifiable? 75
H. Denk
       Morphological investigations on operatively, bioptically and
       autoptically obtained human tissues and organs 75
       Human tissue as the starting point for cell culture 76
       Human tissue as the starting material for the production of
       antibodies for diagnostic and therapeutic purposes 77
       Isolation of active substances (eg. hormones) from human tissue 77
       Operations on cadavers 77
       Removal of human tissue: legally and ethically acceptable? 77
       Humans as guinea pigs? 78

What can we learn from cells?
Cell cultures and their limitations 79
H.A. Tritthart

Clinicians turn to rats
From clinical observation to animal experiments 86
F. Lembeck
       From Hippocrates to prostaglandins 87
       Only a balloon probe 87
       A private patient 89
       Angina pectoris 89
       Hidden genetics 90
       A hormone as a poison 90
       Mental relaxation 90

Pointless repetitions?
Gleaning information and animal experiments 91
W.H. Hopff
       The problem of nomenclature 91
       Information flood against intellectual training 92
       Lack of information 93
       The life work of Arthur Stoll 95

The accusation of repeated animal experiments 96
　　　How does it look on the other side? 99
　　　Has giving information any disadvantages? 99

Is toxicology more than just the LD_{50} test?
New methods of *in vitro* toxicology 100
E. Beubler and B. Schmid

From animal to computer
New physical and biochemical methods for replacing animal experiments 105
A. Saria
　　　Chemical and physical systems 105
　　　The computer as a substitute for animal experiments 106

5. SPECIAL AREAS OF RESEARCH 109

Tiny amounts - great effects
Animal experiments in endocrinology 109
H. Kopera
　　　The history of endocrinology 111
　　　Qualitative hormone determination 112
　　　Quantitative determination of hormones 113

The 'lonely' heart
Cardiovascular research 117
H. Juan and G. Raberger
　　　The heart of the intact organism 120

Small enemies - great dangers
Chemotherapeutic research in animals 121
H. Obenaus and J.G. Meingassner
　　　What is chemotherapy? 123
　　　An example: malaria therapy and prophylaxis 124

Safety for the healthy - treatment for the sick
Cancer research 128
R. Schulte-Hermann and W. Paukovitis

The fight against cancer
Manoeuvers with biological and chemical weapons? 136
H.H. Sedlacek
    The animal protection law 137
    The principle of equality and the balancing of rights 138
    Animal experimentation in fundamental research 142
    The prevention (prophylaxis) of cancer 144
    The treatment of tumours 146

A bridge-head for alternative methods
Immunology 153
G. Wick
    Structure and function of the immune system 153
    Immunopathology 156
    Methods supplementary to animal research 158

Protection of the unborn
Prenatal toxicology 164
D. Neubert
    The incidence of 'endogenous' or 'exogenous' factors as
    causes of abnormal prenatal development 164
    Detection or elimination of a prenatal toxic potential 165
    Current approaches to testing for reproduction-toxic effects 165
    Possible alternatives to testing for reproductive-toxic effects 166
    In vitro methods which can act as substitutes in reproduction
    toxicology 167

Reward and punishment
Behavioural research and ethopharmacology 169
F. Lembeck
    Observations on animals 169
    Research 172
    Is ethopharmacology an alternative? 173

Brain research
The nervous system under investigation 174
R. Gamse
    Study of brain function in animal experiments 176
    In vitro methods of brain research 176
    From experimental research to therapy 177

No pain for the animal?
Pain research 179
R. Gamse
    Acute and chronic pain 179
    Investigations on the conscious animal 181
    Alternative methods 183
    Human studies 183

Animals help animals
Animal experiments improve the maintenance, feeding and treatment
of domestic and useful animals 184
D.F. Sharman and M. Holzbauer-Sharman
    Contagious animal diseases and vaccines 186
    Disorders of mineral metabolism 188
    Pharmacology 189
    Abnormal behaviour patterns 191
    Prevention of stress in the maintenance of useful animals 192

Animals teach students
Animal experiments in lectures and in practice 194
F. Lembeck

Sacrificed animals?
Animals in experimental surgery 197
U. Losert
    Animal experiments for the training and education of surgeons 199
    Animal experiments for material testing 199
    Animal experiments to improve current forms of diagnosis
    and therapy 200
    Animal research in the development of treatment for previously
    untreatable diseases 201
    Animal experiments for the replacement of diseased or
    defective organs 203
    Animal research to explain pathophysiological processes 204

6. PROVISION FOR RESEARCH WITH ANIMALS 206

Is animal experimentation still necessary? 206
P. Skrabanek

The question of responsibility
Ethical aspects and legal requirements 212
N. Zacherl

Suitable or not?
Modern maintenance of research animals 218
H. Juan
       Demands from the animal protectionists' point of view 218
       Requirements from the animal researchers' point of view 220
       Utilisation and 'expenditure' of laboratory animals 221
       Requirements of animal maintenance 221
       Quality of (research) animal maintenance 224

Demand for existing alternative methods
Promotion of new alternative methods 226
F. Lembeck
       Man and animals 226
       Analysis of the current situation 227
       The demand for alternative methods 228
       Promotion of alternative methods 229
       Old aims - new methods 229

Glossary 231
A. Bucsics

Index 242

Introduction

F. Lembeck

The incentive for this book came from the Federal Minister for Science and Research at the time, Herr Dr. Heinz Fischer, and was immediately taken up by the President of the Austrian Academy of Sciences and the present Minister of Science, Herr Dr. Hans Tuppy. The execution of this project was assigned to me.

The demand for 'alternative methods' arose from the widely publicised debate on animal experimentation. The concept of 'alternative methods' became the catchphrase by which any different method was understood. In many ways, the demand for alternative methods resulted from the ways in which animal experiments were performed centuries ago, but which had long since been abandoned. Furthermore, the fact was ignored that for decades there has been a continuous transfer to more modern, and therefore 'alternative' investigative procedures, a change which now finds a rapid impetus through other methodological bases. As soon as the scientific value of a new process is clearly established, the transition becomes a logical consequence, without however becoming an unscientific 'principle'.

Thanks are due to the Council for International Organizations of Medical Sciences (CIOMS), a general scientific organisation for the World Health Organization, and to UNESCO for a clear definition of 'alternative methods'.

"There are still many areas of biomedical research in which, at least for the foreseeable future, animal experiments remain indispensable. An intact living animal is more than the sum of isolated cells, functions, tissues and organs. In the complete animal there are complex interactions, which cannot be encompassed by biological or non-biological 'alternative' methods. The notion of 'alternatives' is generally understood to be the substitution of living animals by other procedures, by which the number of experimental animals can be reduced or experimental processes can be improved. Amongst experimental procedures which are included in the expression 'alternative' are both non-biological and biological methods. To the non-biological methods belong mathematical models of structure/effect relationship based on physico-chemical properties of drugs or other chemicals, or computer models for biological processes. Considered as biological methods are the use of microorganisms, *in vitro* preparations (cell fractions, isolated tissue, isolated viable organs, cell and organ cultures) and for certain requirements, lower animals and embryos of vertebrates. In addition to the experimental procedures, retrospective and prospective epidemiological research in humans and in animals acquire greater importance.

The transition from alternative methods must be seen as complementary to the use of intact animals, and their development, like the use of such methods, should be offered on scientific and humanitarian grounds."

The aim of the present book is to give to the interested, but in the narrower sense, non-specialised reader, a thorough insight into possible alternative methods. To this end it was necessary, in the introductory contributions, to clarify the role of the animal experiment in modern medical science. Thus, it can be made clear how closely modern animal experimentation is associated with both basic and clinical research. From this position it is then possible to discuss the existing alternative methods, as well as their possible further development.

In order to best illustrate the present status, 30 scientists from Austria and other countries and different scientific fields were invited to contribute to this book. The presentation by different experts gives an insight into the many aspects of these methods, specifically based on the present state of research. To aid understanding, all the authors tried to avoid the unintelligible jargon which is commonly used in their specialist fields. A glossary has been appended to clarify unavoidable specialised terminology.

It was vital that the authors of individual contributions should be given enough time to write. Only a contribution written at leisure can offer the reader a format which facilitates easy reading. The editor is grateful for the efforts of all the scientists participating in this project.

References
International Guiding Principles for Biomedical Research Involving Animals, CIOMS, Geneva 1985.

Introduction

The following works appeared while contributions for this book were being prepared. They offer excellent and comprehensive information about the current status and future development of 'alternative methods'.

Alternativen zu Tierexperimenten, ein halbjährliches periodikum herausgegeben von wissenschaftlern der schweizerischen hochschulen und vom fonds für versuchstierfreie. Forschung CH-8032 Zurich, Biberlinstrasse 5.

Alternatives to aninal use in research, testing and education. U.S. Congress Office of Technology Assessment, Washington D.C. 1986.

Archibald, J., Ditchfield, J. and Rowsell, H.C. The contribution of laboratory animal science to the welfare of man and animals. 8th ICLAS/CALAS Symposium Vancouver 1983. G. Fischer, Stuttgart 1985.

Fox, M.A. The case for animal experimentation. University of California Press, Berkeley 1986.

Goldberg and Frazier. Alternatives to animals in toxicology testing. Scientific American 2, 261, 1989.

Paton, W. Man and mouse: Animals in medical research. Oxford University Press, Oxford 1984.

Thelestam, A. and Gunnarsson, A. The ethics of animal experimentation. Acta Physiol. Scand. 128, Suppl. 554 (1986).

Ulrich, K.J. and Creutzfeldt, O.D. Gesundheit und tierschutz. Weissenschaftler melden sich zu wort. Econ, Düsseldorf 1985.

1

The Basis of Animal Experimentation

This section should show how the medical profession began using animal experiments. There are many examples to illustrate that animal experimentation represents only a part of medical research. Just as one may ride a horse to travel more quickly, medical researchers use experimental animals to gain knowledge which has been unobtainable from humans. Whoever undergoes a diagnosis or receives treatment, should realise that animal experimentation has been necessary for their development. However, every doctor who uses animal research will readily adopt alternative methods if they contribute to solving an unresolved medical problem better, more cheaply and more quickly than animal experiments.

From discovery to research
From observations in man to research on animals - an historical review
F. Lembeck

To understand what drives a researcher into a modern scientific or medical field, one can either watch him in his laboratory or read his publications. Either may be equally bewildering and barely intelligible - not just for the non-specialised layman - but also for specialists in other areas of research. Almost all researchers lack the talent to translate their particular laboratory jargon into generally comprehensible concepts. Only specialist journalists are fully conversant with the art of making a generally understandable presentation out of the 'Greek' of a specialized headline. Since most modern research is financed by taxation, the taxpayer has a right to be informed about the significance of such research.

The results of some particularly expensive research projects are understood by only a few people. What does the newspaper mean by reporting that "success has been achieved in a gigantic storage ring in accumulating electrons (e^-) and positrons (e^+) each with 3.5 billion electron volts (GeV), in order to allow them to collide head on?" The high energy and financial outlay involved in such projects is probably necessary to secure an appropriate form of energy production in the future. We tolerate these high costs mainly because we trust the experts who consider it to be money well spent.

The objectives of medical research are close to the interests of the layman because concepts such as risk of cancer, environmental damage, allergies, wasting diseases and genetic defects have a much more personal significance for him. However, what is incomprehensible to him is the significance of an *individual* research project. Whenever I seek to explain that we are able to block specific nerve fibres, and thus the pain filaments in rats by chemical means, and therefore have the potential to study these nerves, I am almost always asked whether a new pain-killing agent, e.g. for biliary colic might result. My answer, that biliary colic is caused by stones which require surgical removal, and that any new selective treatment for pain is still some years away, is of little comfort to the questioner. Perhaps this answer fails to explain that we are primarily only aiming for further insight into the function of the pain nerves without looking directly to the immediate therapeutic benefits.

A visit to the laboratory is interesting and complicated, while inspection of the published results is sobering and not very informative. Nevertheless, the methods of a modern medical researcher are better explained by his publications. Indeed, experimental discoveries, whether in animal research or by other methods, would be valueless if they were not published. By publication we mean the printed scientific paper in a specialist journal, which is freely available on subscription or loan from a university library. This implies that none of our results are state or military secrets, contain no hidden hypotheses or predictions, but are generally accessible and are thus open to expert criticism. This is evident from a consideration of the arrangement of a piece of scientific work.

A scientific paper begins with a title which expresses the content of the work, followed by the names of those researchers contributing to the study and the institution in which it was carried out. This is followed by the 'summary' or 'abstract' (specifically for the reader in a hurry) in which the major findings are summarised in an abbreviated style. There are no language problems since all work of importance is published in English.

The 'introduction' then discusses the earlier knowledge on which the new investigations are based, and defines the current state of knowledge. The basis and objective of the experiments in question are then revealed.

In the 'methods' section, the materials and techniques used are described accurately or the appropriate references to established procedures are given. With animal experiments, the animal strain, maintenance and feeding details, and methods of observation and treatment are presented precisely. The statistical methods used and the outcome for the animal are also mentioned.

This is followed by the 'results' section which, with the text, figures and tables describe all the investigations so accurately that another investigator is in a position to verify them.

The 'discussion' is the most important part of the publication, since it is here that the conclusions which the authors draw from their research are presented. What has been proved should be carefully explained, as well as that which remains unclear and needs further elaboration.

Who then has designed this procedure? It was not an individual scientist, but rather it has evolved over the years as an appropriate means of communicating research work and is now followed by all international scientific journals. Who after all then has posed the problem, instigated the experimental work, chosen the methods and thus provided the initiative for the work? None other than the participating researcher on the basis of his own experience and own moral responsibility, and with a goal he has set for himself! No Ministry nor pharmaceutical firm (although it could well be one of their employees) has initiated the project. Therein lies the cause of the highly personal motivation of the expert researcher. This may not be the case however, if a scientific or economic development results from an existing scientific basis. Here relevant work must be coordinated and a time and money plan submitted. Here it is a case of transfer of basic research into a field of applied research.

The freedom of the researcher in basic research thus seems to be limitless. In fact it is somewhat more modest: he must seek financial support for his project from an appropriate research fund, where the methodological and thematic merits of his application are assessed by two anonymous and independent expert consultants (without honorarium!). The risk of the outcome of his work however is up to him. He may however, need to seek approval from other experts, for example to use radioisotopes, to comply with institutional safety regulations and to undertake proper animal maintenance and infection avoidance. When the results are finally obtained and the work is written up, they are sent to the editor of a scientific journal with a 'request for publication'. The editor again appoints two independent and anonymous referees (without honorarium!) who set about the manuscript like two tax inspectors over an income tax return. They ask questions and make changes or comments which must then be considered by the author. When the content of the paper has finally been approved, the manuscript goes to the printer, and after the final corrections ('proof-reading'), becomes the property of the publisher.

This portrayal of the scientific method is necessary to understand the possibilities for alternative methods to animal research. It is clear that there is no obstacle to the development and application of alternative methods as long as their value stands up to scientific scrutiny. The way to newer and better methods is open, since work which involved outdated methods would not be accepted for publication by any specialist journal. No researcher would carry out experiments on dogs or cats if the study was possible on rats or guinea-pigs, because of the financial implications of such a decision. No endocrinologist would carry out hormone determinations using animals if a reliable physicochemical method was also available for the purpose.

Visits to physiological or pharmacological laboratories demonstrate this development. Up to two decades ago, the sight of an anaesthetized dog or cat, accompanied by much recording equipment, was common. The modern laboratory animal however, is the white rat on which almost all investigations, ranging from behavioral studies to circulation or brain research, can be performed. The number of rabbits or cats used has been considerably reduced. Nevertheless, complex circulatory studies in dogs remain quite indispensable as well as clinical research investigations on human volunteers. In many laboratories however, one now only sees the apparatus and not the animals. The whole animal has often been replaced by isolated smooth muscle, the homogenate from liver cells or isolated nerves, on which functions are recorded with the help of custom-built apparatus.

The exploration of medical problems and the awareness that new knowledge can be spread world-wide by other researchers, represents the mainspring for all research work, even if its practical application is still not directly recognisable.

Was the Italian Riva-Rocci, when over 100 years ago he developed the now universal method of measuring blood pressure with his upper-arm cuff, to foresee that today we would have great difficulty in diagnosing hypertension and controlling its treatment without this simple method? Could Paul Ehrlich, in describing the differential staining of the white blood cells suspect that one day this procedure would be indispensable for all practising clinicians? Was Starling, when he studied cardiac function in heart-lung preparations, able to surmise that this was laying the foundations of modern cardiac surgery and even possible transplantation? For the first ECG apparatus a special room was needed; modern 'alternative' equipment is no bigger than a transistor radio. Fifty years ago, who would have been able to foresee that pregnancy could be diagnosed one week after fertilization using a simple urine sample, and that after a few weeks a photograph of the developing child could be produced by ultrasound imaging? In each case, the path to these new advances began with a chain of new pieces of knowledge, from which ever further alternatives to existing methods resulted. For the animal experimenter the current methods are certainly not sacrosanct, since he is devoted to an alternative method if it brings him nearer to the goal of his research. He cannot however, seek alternative methods in animal research, because they almost always stem from special circumstances and knowledge, the value of which has to be verified.

The following example shows how difficult it is to understand the significance of a new basic research result. In 1920, Otto Loewi noted that frog hearts which had been removed from the body and placed in a suitable apparatus continued to beat for hours. He recorded the cardiac activity, which was strengthened or weakened by stimulating certain heart nerves. Through an ingenious idea he was able to show that the nerves which promote heart function produce adrenaline, while those that reduce heart function, release acetylcholine. The nerve fibres accordingly adjust their regulatory commands by releasing minute amounts of stored messenger substances from so-called neurotransmitters. He recognised that this principle of 'humoral transfer of nerve impulses' must apply to all types of nerve, both peripheral and in the central nervous system. Sir Henry Dale and other researchers subsequently demonstrated in many complicated experiments on dogs and cats, that the discovery first made on frog hearts applies in fact to all mammalian nerves. In 1936 Loewi and Dale were awarded the Nobel Prize for their discoveries. When Otto Loewi achieved his first positive result he was visited by an uncle, a Frankfurt banker, and he privately showed him, full of enthusiasm, an experiment on a frog heart. The uncle however was not at all impressed and put the question: "Tell me Otto, is this really an occupation for a grown man?" (Lembeck and Giere 1968). But how could the uncle have realised that, with this simple method, he could actually *see* a physiological process which also took place in his own body a billion times every second? How could he have suspected that with this finding, the mechanisms of action of existing drugs would be elucidated, opening the way to numerous further drug developments? This episode should only serve as an example. All researchers, particularly those who use animal experiments, know there is only limited knowledge about their work outside a relatively small circle. They may consider themselves fortunate if a generally appreciable useful application results from their work and they themselves live to witness it. In practice, they may only hope for appreciation of their fine work and clear goals.

Reference
Lembeck, F. and Giere, W. Otto Loewi - ein Lebensbild in dokumenten. Springer, Berlin 1968.

From research to cure
The scientific basis of medicine
F. Lembeck

The foregoing contribution should underline the significance attached to the investigation of biological processes. Animal experiments are necessary to gain an insight into these processes through living organisms. Physical and chemical methods make it possible to measure physiological functions, and the modes of action of drugs and toxic substances can be analysed from their effect on the living organism.

Behind all this pressure for knowledge, treatment of the sick and prophylaxis for the healthy stand out as the main object of animal research. Nevertheless, just as the mountaineer dares not dream of sunrise on the peaks, but must concentrate on the sureness of his steps, so must the experimental medical researcher consider the value and reliability of his experiments. As the surgeon must be sure of the performance and outcome of an operation, so the medical researcher should critically view his work on animal experimentation and its intrinsic value. While for the clinician a healthy patient is important, for the medical researcher it is his contribution to the advance of medical science.

However variably, slowly and unexpectedly - and sometimes through luck - a new therapeutic application does emerge from research results. The following examples will illustrate how knowledge resulting from animal experiments may lead to new treatments.

Neurochemistry and Parkinson's disease
A few years ago, following research into neurotransmitters in peripheral nerves, these substances were also detected in specific areas of the brain (see p. 8). This permitted the discovery of nerve fibres which deliver these neurotransmitters to certain parts of the brain. Of particular primary interest was the presence of noradrenaline and it transpired that another, chemically similar substance, termed dopamine, was found in the brain. It finally fell to the Swedish scientist Karlsson to histologically demonstrate these nerve fibres in the brain. Blaschko and Hornykiewicz studied the synthesis and breakdown of these substances in the brains of experimental animals. This was of course a piece of research with no *therapeutic* goal. It was soon found that dopamine-containing nerves were particularly abundant in those areas of the brain which are abnormal in Parkinson's disease. A collaborative study between Birkmaier and Hornykiewicz of the brains of patients who had died of Parkinson's disease discovered a reduction of dopamine, and established that the function of dopamine-containing nerve cells must be disturbed in this condition. An important insight into the pathophysiological mechanism was thus obtained. Meanwhile however, it was also discovered that dopamine was formed in these nerves from

l-dihydroxyphenylalanine (l-DOPA). It was further demonstrated that l-DOPA, unlike dopamine, passes into the brain from the blood. On the basis of this discovery, an experiment was performed in which l-DOPA was administered to a Parkinson's patient, before and during a test of his handwriting. Initially, writing was almost impossible due to the Parkinson's disease, but a few minutes after the start of the l-DOPA infusion his writing became completely normal. Thus, the way was opened to the present day treatment of Parkinson's disease, albeit only after many further stages of clinical research.

Playing with histamine

For many years histamine was a favourite plaything of the experimental pharmacologist. It contracts the intestinal muscles and dilates certain vessels. Histamine can be detected by these biological reactions in the mast cells of tissues, in certain blood cells, and in the intestinal tract. It was further discovered that histamine is formed from certain bacteria and is also present in stinging nettles. The rash caused by touching a nettle is a typical histamine effect familiar to all of us. Medical interest in histamine revealed that the acute allergic reaction resulting from antigen-antibody reaction on the surface of mast cells is associated with violent release of their total histamine content. This release reaction produces the toxic effect known as 'anaphylactic shock', such as can also be precipitated by hypersensitivity to bee stings or certain drugs. It can thus be recognised that certain allergic reactions in the body release the toxic substance histamine, which is stored in mast cells until an acute allergic reaction occurs.

A therapeutic application was revealed by the discovery of specific antagonists to histamine. These substances block the actions of histamine in blood vessels and smooth muscle cells. The use of these anti-histamines in certain allergic conditions is now a routine treatment.

One day, a pregnant woman suffering from an allergy used an anti-histamine. Surprisingly, the morning sickness from which she was also suffering ceased. From this fortuitous discovery arose the development of modern anti-travel sickness agents.

The first anti-histamines had an unacceptable side effect in that they also rendered the patient tired and confused. These sedative side effects interested anaesthetists who sought improved anaesthetic procedures. It was easy to see from animal experiments which of the anti-histamines known at that time were particularly strong sedatives. The first animal experiments had already shown that the effect of each of the sleeping pills, pain killers and narcotics could be clearly differentiated. It was thus possible to develop a special form of narcotic agent which is still used today. However, it was mainly through the treatment of psychoses that the special significance of these 'anti-histamines' was revealed. From this resulted, again through animal experiments, the development of specific antipsychotic drugs which possess almost no anti-histamine effect. Indeed, nothing has changed the image of psychiatry in recent years more than the use of anti-psychotic and anti-depressant agents.

Another experimental observation was no less significant: anti-histamines were able to block all histamine effects, with the exception of the stimulation of gastric acid production. This property of histamine was recognised not only in the animal, but also in man, and led to the use of histamine to test the stomach's acid-secreting capability in man. Much later, the English pharmacologist Black was the first to discover a new anti-histamine which was able to selectively block the action of histamine on gastric acid production. These animal experiments triggered a search for chemically-similar, but considerably more effective, analogues. These substances have yielded the modern drugs by which gastric and duodenal ulcers may be cured, thus sparing patients pain and possible surgery.

Substitute for morphia
In 1934, compounds were tested in the Hoechst pharmacological laboratories which had a chemical similarity to known spasmolytics, such as had been used for gastric, biliary or renal colic. The spasmolytic. i.e. cramp-causing effect could be simply measured on isolated pieces of gut. In order to use the appropriate dose on the isolated gut, the pharmacologist Schaumann initially tested the tolerance of these new compounds on the whole animal. He injected mice with the compounds at different doses, i.e. he made what is today decried as a lethal dose (LD_{50}) determination (see p. 100). Thus he discovered in one of these compounds, in one dose and without any toxic effect, a symptom which had hitherto only been observed with morphia. The injection of a small dose of morphia led to a peculiar S-formed curvature of the mouse's tail. Could this new substance possess not only spasmolytic, but also a morphine-type action? Conclusive experimental studies on rats, rabbits and dogs showed all the effects also found with morphia, ie. on pain, respiration and body temperature. Thus, the first synthetic morphia-like analgesic was discovered which would soon be applied clinically and which is still widely used today.

Clover poison - prophylaxis for thrombosis
In North Dakota in 1929 numerous cattle bled to death after surgical castration or de-horning. It was quickly suspected that rotting clover contained a poison which gave rise to this blood coagulation disorder. As clover feeding could not easily be discontinued, there was a determination to discover the cause of the poisoning. The biochemist Quick found that the haemorrhage was due to a specific blood coagulation disorder, namely a deficiency of the clotting protein known as prothrombin. He subsequently developed a special test for this - the 'prothrombin time' - which has since become a routine test in every clinical laboratory.

Quick administered chemical fractions from the rotting clover to experimental animals and was then able to show in which fractions the toxin was concentrated. Finally he succeeded in isolating the toxic agent and in proving that this substance inhibits the formation of prothrombin in the liver.

A few years later it was clinically recognised that a reduction of blood coagulation capability confers an important protection against the thrombosis which can occur after surgery or cardiac infarction. The ability of the blood to clot can also be reduced by heparin, but as heparin must be injected it is not suitable for long-term therapy. However, the 'rotting clover toxin' and similar synthesised compounds (coumarin derivatives) can reduce blood coagulability following *oral* administration. Serial Quick prothrombin time tests are important in thrombosis prophylaxis in achieving the correct dosage, and thus in maintaining a balance between adequate haemostasis for slight wounds, adequate anticoagulation and absence of severe bleeding.

However, another use for coumarin should also be mentioned. Animal research soon revealed that the inhibition of blood coagulation by coumarin derivatives varies considerably between different animal species; rats are by far the most sensitive. On this basis, coumarin derivatives are not only the best but also the least dangerous rat poisons to humans. Rats ingest the poison and subsequently bleed to death from bites or other wounds. The danger to humans or pets however is very slight, and even in cases of poisoning, the treatment is straightforward since vitamin K is an effective antidote. For this reason all earlier rat poisions which are also very toxic to humans have been discontinued. The use of coumarin as a rat poison is the one LD_{100} test which is not only sanctioned for the non-medical user without his having to seek legal approval, but which is even prescribed by pest control regulations.

Sick animals as experimental tools?
Experimental diseases in the animal as a model for therapeutic research
F. Lembeck

Parts XVI/1-25 of the *Handbuch der Pharmakologie* bear the title "Producing disease conditions experimentally". These issues contain much which could shock the anti-vivisectionist, although they also review many alternative methods developed over the course of time. The experiments consisted of feeding animals on a vitamin-free diet in order to be able to subsequently test the effect of vitamins, since no chemical methods of detection existed. Hormone-producing organs were removed as the only way to discover new hormones. Epilepsy-type convulsions were produced in order to develop anti-epileptic drugs almost free of side effects, and animals were infected to test new chemotherapeutic agents. Tumours were produced by chemical agents or cell transfer to test the effect of 'cancer agents' or to exclude such effects. Are all these methods necessary? Must methods of this kind continue to be used? The following examples, old and new, should help to answer these questions.

Vitamin research
Beri-beri was a widespread disease in south east Asia which led to death from severe neurological abnormalities, heart failure and gastric disorders after prolonged suffering. In 1890, the Dutch doctor Eijkman, working in Java, noted that the hens in his hospital which were fed on scraps - mainly polished rice - developed paralyses similiar to those seen in humans with beri-beri. Feeding the hens with unpolished rice cured this condition. Thus it was realised that the husk of the rice grain must contain a substance, lack of which leads to beri-beri. Clinical investigations subsequently confirmed his observations in hens. Thus, vitamin B_1 was discovered, and for several years vitamin B_1-deficient pigeons were used for the biological assay of vitamin B_1 in foodstuffs. Synthesis of vitamin B_1 was first achieved in 1936. Today its production is cheap and it is possible to measure it chemically. Animals are therefore no longer needed for the detection of vitamin B_1.

Has any reader seen a small child with fully developed rickets? Who is still aware that in the famine years after the First World War American doctors came to Vienna to study the numerous cases of starvation oedema and rickets? Who still remembers the use of ultra violet radiation, by means of which vitamin D, lacking in rickets, is produced from sterols? When rickets was first successfully produced in rats by feeding an appropriate diet, it became possible to give the previously introduced prophylaxis with cod-liver oil a more secure biochemical foundation. The optimum dosage was first assessed in this way, which was important, since high doses of vitamin D have toxic effects. Today all this is history since synthetic vitamin D is routinely and prophylactically given to all infants.

Pernicious anaemia is a disorder of red blood cell production; the adjective 'pernicious' describing its virulent clinical course. On the basis of animal experiments, which actually had nothing to do with this condition, it was discovered that it was due to failure to absorb a specific factor in food. As there was no experimental model for this disease, liver extracts, which contain this factor, could only be tested by their clinical effect. Then however, came unexpected help from a microbiological study into the essential nutrients of the specific bacteria used in cheese-making. It was discovered that two of these bacterial nutrients were also present in the liver extracts active in pernicious anaemia. Thus, through microbiological methods, which can be performed quickly and cheaply, an elegant technique became available which could detect and measure these factors. One of the factors is cyanocobalamin (vitamin B_{12}), lack of which leads to pernicious anaemia, the other is the vitamin folic acid, lack of which gives rise to another form of anaemia.

Later it was found that certain moulds of the *Streptomyces* species produce not just valuable antibiotics, but also considerable quantities of vitamin B_{12}. Thus, vitamin B_{12} could be obtained cheaply and in any amount required. Folic acid is also used by some microorganisms and malaria-causing *Plasmodia* as an essential metabolic building block. The development of folic acid antagonists was the starting point for the discovery of effective anti-malarial drugs. It is clear that in this area, microbiologists have offered some very important alternatives to animal research, which are now widely used in modern drug development.

Hormones
Insulin is an essential drug in the treatment of diabetes, and was discovered through animal research. It is obtained from the pancreas of slaughtered animals, and its activity is determined biologically in animal experiments.

Between the discovery that in the dog, removal of the pancreas results in a rise in blood sugar similar to that seen in human diabetics, and the discovery of insulin, there was a gap of several decades. Oral administration of pancreatic extracts had no effect, although the injection of extracts was at least free of serious side effects. The discovery of the essential hormone insulin for the control of blood sugar levels came in the summer of 1921 in Toronto. A quick decision by a clinician and a medical student, to experiment in the laboratory during the summer vacation, together with the help of an experimental biochemist and the interested director of the institution, created a successful research team. No application for a research project, no long wait for an animal licence for dogs which had been rendered severely diabetic through pancreatectomy, and with a restricted survival capacity, delayed the initiative. Even today there would be no other way of discovering insulin - certainly none so quick and conclusive!

Insulin-containing extracts from the pancreas were soon used for substitution therapy in diabetes. There were no questions about side effects of the still largely unrefined extract since it was a case of maintaining life. But how should the amount of insulin be measured, when there was still no clue to its complicated structure? The drop in blood sugar due to insulin was studied in rabbits, a method which first established the requirement for an internationally recognised standard preparation (see p. 113). This method of determination is no longer used in rabbits for establishing a glucose profile in patients. Today the concentration of injectable insulin can be measured by radioimmunoassay, although this method cannot differentiate between insulin preparations with rapid or delayed effects.

Chemotherapy

The discovery of penicillin is probably known to many readers. The observation that certain species of mould yield a substance which destroys bacterial cultures was made *in vitro* and thus by an 'alternative' method. This method is still widely used today to find new effective chemotherapeutic agents. After sufficient amounts of penicillin had successfully been extracted from mould cultures it was possible, in animal experiments, to establish the safe dose of penicillin. These experiments also revealed a property of penicillin which it would not have been possible to show in tissue culture. Penicillin penetrates into most tissues, but fortunately not into the brain. If it is injected experimentally into the brain it leads to severe convulsions similar to to those seen following strychnine administration. Thus, if penicillin passed from the blood stream into the brain, then in spite of its demonstrable antibiotic efficacy, it would be totally useless therapeutically. Under the pressure of the situation created by the war, penicillin, after some preliminary investigations on animals, was finally therapeutically tested on humans.

The experimental bases of chemotherapy were studied 40 years before the introduction of pencillin by Paul Ehrlich. Around the end of the century, *Treponema pallidum* was found to be the pathogen responsible for syphilis. Attempts to grow similar bacteria in *in vivo* culture systems were unsuccessful, but rabbits could be infected by *Treponema pallidum*. At this time it was already known that organic arsenic compounds were effective in another animal infection caused by protozoa. On this basis Ehrlich used rabbits infected with *Treponema* to test newly developed arsenic derivatives. These arsenic compounds were extraordinarily toxic, giving rise to three questions:

- What dose would the rabbit tolerate?
- What dose would destroy the *Treponema* in rabbits?
- How great is the gap between the lethal dose of an arsenic compound for the rabbit and for the pathogen which causes the infection?

The compound 'Ehrlich 606' gave the best 'therapeutic index', ie. the best quotients between the LD_{50} for the rabbit and the LD_{50} for *Treponema* (see p. 100). Without preliminary studies into chronic toxicity, toxic effects on the liver or embryo, or the possibility of carcinogenic effects, clinical trials of this compound were largely performed on the patient. However, of what importance were even significant side effects where there was the prospect of also curing syphilis in man? Without the artificially-infected rabbit, modern chemotherapy would have never have begun.

Experimental animal research into anti-malarial agents, which, from the quinine produced in plants, led to so many effective synthetic drugs, provided the basis for the subsequent development of chemotherapy for bacterial infections. Domagk infected mice with *streptococci* which, by the development of a purulent peritonitis, led to death in a few days. He then tested several dyes as a treatment for this condition, one of which was credited with a 'disinfectant' effect. Prontosil, a bright red dye, was shown to be highly effective and totally non-toxic. All the mice survived the infection without any toxic effects. Only later was it established that there had been no alternative to this experiment, since the bacteria, which were destroyed by prontosil in the infected animal, could not be killed by prontosil in *in vitro* cultures. Eventually it was shown that the chemotherapeutically efficacious derivative sulphanilamide was cleaved from prontosil in the organism, giving rise to an effective anti-bacterial agent. Who could have predicted this? The first use of prontosil in humans was a little known but very dramatic event. Domagk's small daughter developed a severe infection of the arm following an injury with a splinter of wood. Immediate amputation was seen as the only life-saving measure, and a decision had to be made within hours. Should prontosil be used, since a prontosil-sensitive pathogen of the infection was suspected? No preliminary investigations of possible toxic side effects of the compound had yet been performed. The decision was Domagk's alone. He gave prontosil and the arm - and perhaps even the life of his daughter - was saved. Had there been any alternative to this decision? Who would be ready for such a step today to therapeutically use a derivative from a totally new chemical group which was neither officially registered nor clinically proved?

Genetic defects in animals
Other models of disease are not produced by the researcher, but by nature through genetic defects. However, these could be isolated in laboratories, after man had learned to breed them in carefully controlled animal farms. Such defects include rats with spontaneously occurring hypertension, others with a lack of vasopressin (a hypophyseal hormone), and mice with morbid obesity, with specific immunological defects or a high incidence of certain cancers. Genetically-diseased animal species have since been specially bred and represent the best models in which to study the treatment of relevant human diseases.

Allergies
Chance, as well as breeding, has provided an important animal model of disease. An example of this was Sir Henry Dale's use of the contraction of guinea-pig uterine muscle for the biological determination of histamine. The guinea-pigs were given to him by the nearby Hygiene Institute, since they had been used to test anti-diphtheria horse serum and could therefore not be used for further immunological tests. It is clear that efforts were always made to keep the number of experimental animals as small as possible. Only by using these sensitised animals was it discovered that, in acute allergic reaction, huge amounts of histamine are released from the tissues. It is impossible to calculate today how many lives have since been saved because of this discovery. Through these animals the mechanism of the acute anaphylactic reaction was understood for the first time.

2

Drug Development

In earlier times, doctors saw only the symptoms of a disease, e.g. fever or diarrhoea, and tried - as the only obvious possibility - to suppress them with drugs. They then learned to recognise the *causes* of the symptoms observed. Thus, drugs could be developed which were directed against the causes, such as an infection, a hormone deficiency or a blood coagulation disorder. From this resulted many modern diagnostic and therapeutic measures, which are incorrectly regarded only as 'reparative'. Yet these are indeed the scientifically demonstrable, and thus the biologically efficacious component of therapy, combined with the medical and psychological care of the patient which should not be overlooked.

These new diagnostic and therapeutic weapons could only be used after appropriate scientific assessment, and for this purpose the animal experiment is still irreplaceable. However, this should no longer be regarded as vivisection as performed at the start of experimental medicine in the 19th century. Progress has been made from one alternative to another without neglecting the overall picture.

The clinical outlook
Discovery of drugs through clinical observations
F. Lembeck

"Research into drug effects began when the first doctor treated his first patient and observed the result, since all doctors conduct their research continuously. No two patients are quite alike and the doctor must determine, by experiment, which treatment best suits the individual patient. The experienced doctor's opinion is often the best guide to the treatment of the individual patient, but it seldom yields satisfactory evidence of the value of new drugs. Therefore research is often undertaken to obtain objective evidence of the effects of new drugs and to unequivocally establish their value via well planned full-scale experiments" (Gaddum, 1953).

"This drug fulfils all safety conditions for your cat's treatment, since it has been tested for years on humans", is now commonly seen in leaflets extolling veterinary-medical preparations. This is true if the preparation has had extensive therapeutic use in humans, and it has also been conclusively shown to be effective in the treatment of a feline disease. Indeed, it is not exceptional that many modern drugs are used on both humans and animals.

In fact, man has often also been the first 'experimental animal'. Some drugs still in use today had actually been discovered in the earliest historical times. These early discoveries were achieved by the 'sucking method', i.e. people put in their mouths, tasted or ate, everything they found in the way of fruit, leaves or roots. Thus the term 'deadly nightshade' clearly expresses the toxic effect resulting from the ingestion of these fruits. The purgative effect of the oil pressed from the castor oil plant was easily discovered, but by what chance could man have discovered the effect of ergot in strengthening labour pains? It should be noted that in old medical books, one occasionally finds descriptions of things which, despite their attractive colour, unusual taste or smell, are either ineffective or poisonous, the dangers of which could not have been foreseen.

People have used arsenic from early times as a tonic (stimulant); only in this century was its relative ineffectiveness and carcinogenicity discovered. Calomel, a mercury compound, was used as a mild paediatric laxative for generations until Faehr described the serious ill effects of its chronic use. These symptoms are similar to those seen with other mercury derivatives, and which later became known as 'Minimata disease'. Since they stem from nature, and thus from 'God's own apothecary', not only the ineffectiveness, but also the danger of some 'natural remedies' have been overlooked by man. We have only modern animal research to thank for the fact that it has been possible to discover the damaging effects of many medicines which had been previously used for years.

However, the discovery of drug effects on man is not only an historical event. Even in the early days of the systematic development of medicines, including animal experiments, drug testing was carried out. It will be shown by the following examples that it was usually an unexpected observation, and that the right observer was present at the right time and carried out the necessary 'clinical observation'.

The beginning of anaesthetics
About 1840 a travelling showman was touring the United States. He erected a simple chemical apparatus on his stand with which he was able to produce laughing gas. He allowed the public to freely breath the laughing gas at his stall, and the onlookers were

delighted at the comical activities and uncontrolled expressions of these half-anaesthetized people. Imagine today a magician anaesthetizing rabbits and allowing them to reel around in a similar way. The audience would not be at all amused; indeed they would probably be rather indignant. In 1844 the dentist Horace Wells was by chance present at this exhibition. The still slightly dazed victim stumbled over the fist row of seats, injuring himself, but did not complain of any pain. This observation gave Dr. Wells the idea of extracting a patient's tooth after inhaling laughing gas, and paved the way for the development of modern anaesthesia!

Heart drugs

In 1780, the action of digitalis (the name generally used for glycosides affecting the heart) was discovered in the run-down Birmingham practice of Dr. William Withering under conditions which we can hardly imagine today. He was treating numerous wretched patients, although there were at least some effective drugs available. The circumstances in the practice probably exceeded anything we now see from the disaster areas of the Third World. During this massive undertaking, Dr. Withering heard of a tea mixture which helped with dropsy (mainly oedema caused by heart failure), and made life tolerable again. He found that the effective component of the mixture could only be foxglove and studied its action. He was soon aware of the narrow gap between its therapeutic and toxic effects, but quickly saw the 'clinical implication' of this agent. It was the start of a new therapy, although to reach the present status of clinical therapy with refined glycosides, the experimental and clinical developments lasted for generations (Erdmann, Greeff and Skou 1986).

Reduction in blood sugar

The discovery of Prontosil, the first chemotherapeutically effective sulphonamide, was described on p. 16. Subsequent modifications to its structure produced derivatives which proved to be considerably more effective. The clinical desire for an injectable drug was pursued and a derivative suitable for this purpose was finally released for clinical trial. However, the clinical tolerance of this preparation was disappointing, since the patients became restless and suffered from outbreaks of sweating shortly after the injection. From the 'clinical viewpoint' (Franke and Fuchs 1955), these symptoms resembled those of hypoglycaemia, such as can be observed following insulin overdose. Blood sugar determinations revealed the accuracy of this suspicion, and confirmed that this sulphonamide led to a fall in blood sugar; the symptoms could be relieved by administering glucose. This was the discovery of a new potential treatment of diabetes, although once again, its present status was only achieved after numerous further animal and clinical studies.

Treatment of hypertension
The discovery of the effect of clonidine (Catapresan) on blood sugar levels been extensively borne out by animal experimental and clinical research. The compound originates from a chemical group with a vasoconstrictive action. It was found that clonidine penetrates the skin very easily and contracts the muscles around hair follicles, thus making the hairs stand erect. Erect hairs are more easily cut short by the electric razor and thus clonidine was used as a 'pre-shave' additive. Later electric razors were more efficient however, and pre-shave lotions became superfluous. Then the use of clonidine for colds and nasal congestion was considered, because vasoconstriction shrinks the inflamed nasal mucous membrane and makes breathing easier. Around this time, a vial of concentrated clonidine solution stood on the desk of the pharmacologist who was investigating this use for clonidine and who by coincidence also had a cold. Seizing the opportunity, he dropped a quantity of the solution into his nose. He was later found sleeping deeply at his desk and 'clinical instinct' led others to believe that clonidine could have a soporific effect, To check this however, a suitable dose had first to be established in appropriate animal experiments. It was demonstrated primarily in studies on the anaesthetized dog that clonidine, by affecting central circulatory regulation, leads to a long-term drop in blood pressure. These studies led to its clinical use in treating hypertension (Kobinger, 1984). Thus it is clear that the use of animal experiments and clinical trials led to success because these were the only ways in which erroneous applications could be excluded.

Contergan
A young married couple sat next to me on a short train journey. He had deformed hands and forearms, like those of Contergan victims, and his wife helped him with anything he could not do himself. This reminded me of how much time had elapsed since this shocking tragedy and how sustained the consequences are, since his mother had used Contergan as an apparently harmless sleeping pill in the first months of her pregnancy. How could these malformations have occurred when Contergan had shown no such effects in preliminary animal experiments? Since then, new and pertinent tests have been developed and must now be undertaken on all newly developed drugs. Nevertheless, it is comforting to be able to say that no further abnormalities of this kind, caused by other medicaments given during pregnancy, have occurred since then.

But let us not forget how the connections between taking Contergan and the appearances of deformities were discovered. At a meeting of paediatricians, mention was made of a small number of cases of unusual malformations in neonates. This aroused the 'clinical instinct' of other delegates who had seen quite similar and separately observed cases. The correlation of these individual observations soon led to the common denominator, namely the administration of Contergan.

References

Erdmann, E.K., Greef, K. and Skou, J.C. Cardiac glycosides 1785-1985. Biochemistry-Pharmacology-Clinical relevance. Boehringer Mannheim International Symposia. Steinkopff, Darmstadt 1986.

Franke, H. and Fuchs, J. Ein neues antidiabetisches Prinzip. Dtsch. med. Wschr. 40 (1955) 1449-1452.

Gaddum, J.H. Über klinische erprobung von neuen Arzneimitteln. Wein. klin. Wschr. 16 (1953) 297-301.

Kobinger, W. Central anti-hypertensives. Haemodynamics, Hormones & Inflammation 2 (1984) 107-123.

The unexpected result
Findings in animal research: the key to the understanding of function, diagnosis and therapy
U. Holtzer

In connection with important discoveries in research, it is often remarked that the significance of animal experimentation is over-estimated; in some cases the value of animal experiments for medicine and the natural sciences, even of those performed in tha past, is completely discounted. In reality, most of our current biological and medical knowledge is the result of planned research which has also included animal experimentation. Furthermore, purely random observations, which had nothing to do with particular experiments, have also led to important advances in the understanding of the human organism and the potential for the treatment of diseases. The occurrence and scientific significance of such unplanned, and therefore unexpected, results of research with animal experiments will be discussed in this section.

In 1904, Henry H. Dale worked in the Research Institute of Burroughs-Wellcome and Company on the pharmacological basis of ergot. One day he received an adrenal extract from the chemists in order to check, as he had frequently done before, how strongly the extract acted in increasing blood pressure. It was already known that these hypertensive effects were attributable to a substance called adrenaline, but there were still no analytical methods for its determination. Dale had performed an experiment on the effect of ergotoxin, an active compound in ergot, on the blood pressure of an anaesthetized cat. At the end of the experiment he decided to test the adrenal extract on this experimental animal as well. To the amazement of the experimenters, and contrary to all earlier findings, the injection of a small quantity of the adrenaline-containing extract led not to a *rise*, but to a *fall* in blood pressure. As this phenomenon was also observed in a subsequent experiment, Dale discounted coincidence and began a precise investigation of this effect. This 'adrenaline conversion' was the first indication that adrenaline acts on two types of receptors in the organism. Thus, on so-called alpha-receptors, adrenaline exerts, amongst other things, a blood pressure-increasing effect, and its action on beta-receptors reduces blood pressure. The latter is first discernible when the adrenaline action on the alpha-receptors is impeded by so-called alpha-blockers, e.g. ergotoxin. Decades of further chemical, physiological and pharmacological research were still needed before this initial observation, which was only possible on a living anaesthetized animal, led to the use of alpha- and beta-blockers in the treatment of heart and circulatory diseases. In similar tests, Dale later discovered the blood pressure-reducing action of histamine, and thus created the important foundations for the investigation of allergic processes.

In the 1940s, W. Paton worked in Oxford on the pharmacological characterization of muscle paralysing substances (biquaternary polymethyl salts) which act like the arrow venom curare. In tests on isolated frog muscle he showed that a compound of this series (decamethonium) prevented this effect. Since an antagonist against muscle relaxants also seemed clinically interesting, this apparent antagonist was also tested on an anaesthetized cat. In addition to the anticipated effect, the researchers observed an unexpected fall in blood pressure. This effect, which for the differentiation of muscular action, could have been observed only in the animal with an intact heart and circulatory system, was confirmed in all subsequent experiments. Its further investigation finally led to the introduction of hexamethonium; the first drug for high blood pressure.

Certainly, research into the regulation of blood pressure makes it clear that with so complex a subject, experimentation on the whole animal cannot be abandoned. Blood pressure does not depend only on the widening or narrowing of blood vessels and on cardiac performance, both of which are controlled by the autonomic nervous system. Superimposed centres in the brain stem play an important, if still not clearly understood, role in this closed-loop control. However, it is not just the nervous system which contributes to blood pressure; a series of hormonal factors also participates. If peripheral blood pressure falls, the kidneys excrete the hormone renin. This in turn splits a peptide from angiotensin I, formed in the liver, which is converted into angiotensin II by an enzyme in the blood. Angiotensin on the one hand narrows the peripheral blood vessels and thus raises the blood pressure, and on the other hand releases aldosterone from the adrenal cortex. This hormone causes the kidneys to retain salt in the body and thus brings about an increase of blood volume and as a result, a rise in blood pressure.

Since high blood pressure is very common, and often leads to death from stroke, cardiac infarction or chronic kidney failure, it is understandable that new and improved ways of treating hypertensive patients are always being sought. Experiments on rats with genetically-determined high blood pressure have brought advances in therapy. Awareness of the role of aldosterone has made the combination of a low-salt diet and diuretics (water elimination agents) an important component of anti-hypertensive therapy. A further advance in the management of previously poorly managed hypertensives was achieved with the introduction of clonidine which lowers blood pressure through the brain stem. Compounds were developed several years ago which inhibit the conversion of angiotensin I into angiotensin II. However, because of differences between patients, and the often differing causes for hypertension, there are still always patients who gain little or no relief from existing drugs. Even earlier it had been discovered that the heart itself produces a hormone which promotes salt excretion, and thus the excretion of urine, thereby widening the blood vessels and lowering blood pressure. To what extent this atrial natriuretic peptide will play a role in the treatment of hypertension depends on further research work which is primarily possible only in an experimental animal.

The charge has very often been levelled that animal experiments are carried out on conscious animals without any pain relief. However, in heart and circulation research for example, the importance of anaesthesia for research has been well demonstrated. Everyone knows that blood pressure rises due to excitement, fear or severe pain. This is predominately due to the release of adrenaline from the adrenal glands as a response to stress. It is easy to appreciate that the complex nervous and hormonal reactions which we commonly designate as stress response are bound to interfere with all test results. Indeed, long before any of the factors involved in stress were known, Carl Ludwig, Professor of Physiology in Leipzig in the second half of the 19th century, had demanded on methodological and ethical grounds that the anaesthetic methods in use since 1850 should always be used in animal experiments (see p. 18).

A considerable amount of biological and medical knowledge has resulted from unforeseen observations. Even the discovery that diabetes was due to a defect of the pancreas was accidental. All these examples should show that - in animal experimentation more often than anywhere else - unexpected results may emerge which are at first bewildering, then interesting, and, in their subsequent pursuit, of world-wide therapeutic importance in this regard. Animal experimentation is no exception: penicillin used in tablet form was only discovered because the manufacturer added a compound, which should have prevented the growth of bacteria, to the crust of a culture mould. However, this substance became incorporated into the penicillin molecule and thereby made the drug resistant to stomach acids.

It does not necessarily have to be the researcher who provides the unusual result. At the end of the 19th century, Professor Ringer in London perfused isolated frog hearts with a sodium chloride solution; it was already known that the perfusing solution should have an adequate salt content. As the laboratory assistant was absent on one occasion, the Professor himself dissolved the salt in distilled water. The test did not work well and only succeeded satisfactorily after the laboratory assistant had returned. He confessed to the Professor that instead of distilled water, he had always used London tap water to prepare the solution for the experiment. In fact, this water contained other ions as well as sodium and chloride, and it has since become established that these are also necessary.

When significant discoveries with animal experiments are mentioned, one increasingly hears an objection that the really important discoveries have already been made, and that no more animal experiments should be used; i.e. that we are now beyond this stage. Unfortunately, it will be a long time before these research methods can be abandoned. The human organism is naturally complex, and many of its functions are still so inadequately explained that no one is yet capable of programming a computer to even remotely simulate the function of our body (see p. 106). In addition, we should not forget the many patients who still cannot be satisfactorily treated. For example, for many nervous system diseases there is still not even a palliative, let alone a causal therapy. Different forms of rheumatic disease can only be partially relieved, but not treated causally. Should we leave in pain thousands of patients who suffer from them, and who are crippled as a result of their illness, and take all hope away from them because the disease mechanism cannot be adequately explained?

Naturally no researcher can be sure that his experiment will furnish the expected result. If this were the case he would not need the experiment in the first place. Although he naturally plans his experiment as well as possible, he must always be prepared for unforeseen eventualities. The quality of the researcher is demonstrated by his reaction to such unpredicted events; whether he, like Dale or Paton, will be able to see an unusual observation, not simply as an experiment gone wrong, but will perceive its special significance as new knowledge and will investigate it thoroughly.
"Chance only touches the minds of those who are prepared".

References
Dale, H.H. Adventures in physiology. Pergamon Press, London 1953.
Issekutz, B. Die geschichte der arzneimittelforschung. Akadémiai Kiadó, Budapest 1971.
Paton, W.D.M. Hexamethonium. Brit. J. clin. Pharm. 13 (1982) 7-14.

Strategic developments
Targeted search for new drugs in animal experiments
R. Kilches

In this century our lives and everything associated with them have borne the stamp of chemistry (Figure 1). Analytical experiments have clarified the chemical structure of even the most complicated molecules in microorganisms, plants or animals (organic chemistry). The synthesis of urea by Wohler 150 years ago was the springboard for this development, since he first produced in the test tube something which only 'nature' was previously thought capable. Meanwhile it was a question of technique and good management as to whether an organic compound was synthesised 'in the test tube' (i.e. in a high technology clinical plant), whether it was synthesised from microorganisms (e.g. vitamin C, penicillin or steroid hormones) or, more recently, produced by genetic technology in which the peptide synthesis mechanism is introduced into a microorganism (e.g. insulin). This variety of synthetic potential has meant that we now know of many more synthetic substances than are present in nature.

Synthetic washing agents, textiles, wrapping foil and tubing, pesticides, herbicides and preservatives are now in everyday use, as are an unlimited number of light- and colour-fast dyes for printing, clothing, colour films and motor cars. It is therefore not surprising that scientists should attempt to synthesise drugs which had previously been isolated from plants. If successful, it may also be possible to synthesise structurally similar compounds. Indeed, this approach has yielded numerous compounds with significant therapeutic effects, and which fall into three groups:

1. A natural substance may serve as the model for the synthesis of analogous compounds, which are cheaper, more effective, less toxic and have fewer side effects. Examples include local anaesthetics synthesised from cocaine, synthetic analgesics from morphia, and the specifically anti-inflammatory dexamethasone from cortisone.

2. There are relatively few cases where synthetic analogues cannot be derived from a natural substance. These include the glycoside cardiac agent digitalis, some hormones and certain antibiotics. In some cases however, the purified natural compound first yielded a specific drug through subsequent partial synthesis. This approach resulted in the specific uterine contracting agent, methylergobasine, bromocryptine, which specifically affects brain structures, as well as penicillins with oral action or a particular antibiotic spectrum.

3. Additionally however, an increasing number of compounds with therapeutic effects have been found, for which there was no model in nature. Examples of these

include modern anaesthetic agents, diuretics, psychopharmacoceutic agents and tuberculostatics.

"We've already tested 605 of these compounds, that's enough for me Professor Ehrlich!"

Figure 1. Screening methods are used to determine, out of a series of chemically similar compounds tested, the one with the highest activity, the fewest side effects and the most appropriate and long-lasting therapeutic action. From these experimental results the scientist can deduce which further compounds promise even better properties. When is the best substance found? This cartoon shows that, for Paul Ehrlich's colleagues, patience ran out at compound 605. About 1910, Paul Ehrlich was screening the organic arsenic compounds for an agent with high activity against the syphilis pathogen, but low toxicity for the infected animal. This was the foundation of modern chemotherapy. He may however, have tested fewer than 606 substances. '606' was probably the sixth compound of the sixth series synthesised. It was subsequently used clinically and was the only effective drug for syphilis for decades, until it was replaced by the even more efficacious and non-toxic penicillin.

The possibility of synthesising not just an active substance but also many chemically similar substances demanded new strategies. Of particular interest were those compounds with the optimum therapeutic effects and the least toxicity. Neither of these qualities can be deduced from the chemical structure, although they may be gleaned from their specific properties, such as water solubility or tissue-binding capability. On the other hand, pharmacological actions cannot be determined without studying the action in appropriate biological systems to find the most effective derivatives. These are called 'screening' tests, and in order to achieve a rapid overview of the drug effect, usually embrace several of the simplest possible methods.

An analogy with dyes illustrates that a new series of beautiful red textile dyes is optically impressive, but the manufacturer and purchaser are interested in quality as well as appearance. They ask whether the colour is fast, withstands boiling, ironing and dry cleaning, is light-fast and does not stain the skin or other textiles. Each of these properties must therefore be tested. It is the same with motor cars. All innovations are individually tested long before a new series is produced; the durability of the brake linings, the temperature tolerance of the lubricant, the life-span of the sparking plugs. With some exceptions, it is no different with the pharmacological properties of analogues of a known drug:

1. Soon after bacteria were discovered it was found that they could be destroyed by some chemicals as well as heat. The first disinfecting agent was phenol. Phenol derivatives had a higher efficacy however, and the 'phenol-index' evolved to quantitate their increased effectiveness compared to the parent compound. Subsequently, newer applications of these compounds were tested, for example in swimming baths, operating theatres or as additives to skin care agents.

2. Screening for antibiotics seems to present a 'never ending story'. Many infectious microorganisms can be grown *in vitro* in culture, and the addition of an antibiotic to the culture medium inhibits micro-organism growth. Using this method, it can be shown which particular species of mould produce much penicillin. After investigating hundreds of thousands of moulds, some were found which yield more than a thousand times the amount of penicillin than the original type, thus allowing much cheaper antibiotic production.

Screening *in vitro* however, is inadequate, for it was mainly due to studies in experimentally infected animals (see p. 15) that it was seen whether the antibiotically active compound is also transferable from the host organisms. The first animal experiments revealed that over 90% of the antibiotic agents effective *in vitro* had to be rejected as too toxic.

The discovery of 'new generations' of antibiotics seems to be inexhaustible. In 1944, Waksman discovered the first antibiotic effective against tuberculosis in specific microorganisms, namely streptomycin. From certain species of these microorganisms numerous further antibiotics were isolated, which today are being used therapeutically. In 1981, a new group was discovered which merits special mention. The avermectines have no antibiotic effect on bacteria, but are active against parasitic worms. These agents represent the first effective and easily applicable therapy for blinding disease (onchocerciasis). Thirty to forty million people suffer from this disease in Central Africa and Central America, and as a consequence, over one million of them become blind. The disease can be cured by taking two tablets containing a few milligrams of the antibiotic at an interval of one year. In animals, a semi-synthetic analogue facilitates the simultaneous treatment of infection due to maggots, mites and ticks.

3. Since new compounds are tested not for their overall action, but for individual pharmacological effects, isolated organs may be extensively used (see p. 68). An isolated intestinal muscle or heart allow effects to be measured, which affect the muscle or the nerves pertaining to it. An action on the peristaltic reflex can be examined on a few centimetres of isolated guinea-pig gut. For some years it has been possible to culture isolated nerve cells, and their response to various drugs can be determined electronically. The specific binding of substances to certain receptors can be measured in small tissue samples using radioactively labelled compounds. If it is discovered *in vivo* that a drug influences an enzyme in a certain way, similar compounds can be selected for their action on the isolated enzyme system.

Every year new and improved 'isolated systems' are coming into use. That *one* specific effect can be rapidly quantified is common to all these screening methods. Thus, a large number of fairly ineffective derivatives can be 'weeded' out as unsuitable at an early stage.

The advantages of these methods are founded on years of experience. Thus the glycoside cardiac agent digitalis can be tested on the hearts of frogs or guinea-pigs, but not on those of toads or rats, since in the latter they are ineffective in biochemical tests. Material from the abattoir (p. 72) or human surgical material (p. 75) is used for certain, but very limited, purposes. The use of the isolated organ or tissue (see p. 56-59 and 83) is still undergoing further development, and is therefore an important alternative to research on the whole animal.

4. Nevertheless, further screening methods demand the use of intact animals (see p. 12 and 123). Modern laboratory technology however, permits many body functions to be measured on anaesthetized rats; measurements which 30 years ago could only be assessed on cats or dogs. The rat is also the main model for studies of the functions of the central nervous system. The action profile of many compounds even enables

rats' behaviour to be defined under specific test conditions (p. 172). Even complicated procedures in the nervous system (p. 174) and also in pain research (p. 179) today use not only mainly the rat, but also highly developed isolated systems, as shown in Figure 17 (p. 180). Many developments arose chiefly through refinements in measurement techniques (p. 105). In this way, as well as through the modern breeding of laboratory animals (p. 218), it is possible to obtain quicker and more reliable results with many fewer research animals. From this it is seen that the 'test pyramid' (see p. 33) used in the modern pharmaceutical industry, can now be constructed with considerably fewer test animals through the numerous stages of basic research.

Since not only efficacy, but also safety is desirable in a drug, screening must be extended to study these aspects. This includes studies of resorption, organ binding, structure and excretion. Microchemical analyses and isotope technology have provided an important breakthrough in supplying sufficient measurable results from relatively few animals. Nevertheless, the total number of animals used for such investigations remains relatively high since consideration must be given to safety in long-term use of a drug, and with daily (p. 44 and 47) and even life-long contact with chemicals.

Finally it should be said that drug development is no longer a case of 'happy-go-lucky' synthesis and experimentation. Network information systems and careful calculations help to reduce time and expenditure, including the number of experimental animals. Previously, a great deal of time and money would have gone into a new drug - typically a development time of 12 years, and an expenditure, from the first indications of its action to the start of marketing, of 4000 man-years and 400 million Swiss francs. These figures show that the test pyramid concept is not needed to advance the development of alternative methods, since every possible alternative method is seized upon and used in modern drug development.

The particular value of this strategy lies not only in achieving breakthroughs in new areas of therapy, but also in improving existing drugs. The chemically pure penicillin derivative of 1986 can no longer be compared with the crude penicillin-containing extract of 1946. The 1986 Volkswagen Golf too now only has the 'VW' on the radiator in common with the 1936 models. The selection of better drugs requires suitable research animals. In this it is like other biological developments. In breeding new fruit and vegetable species not only is the flavour tested, but also the yield, climatic stability, storage properties and fertiliser requirements. In breeding strains of hen, the suitability for meat or egg production must be considered. Each of these strategic developments in the industrial nations, and the subsequent deployment of successful developments, particularly drugs, contributes to the solution of many problems in the Third World countries. The sacrifice of experimental animals probably reduces the misery in developing countries better than many other measures.

From mouse to man
Transferability of the results of animal experiments to man
W. Kobinger

Animal experiments are not transferable
It is often said by opponents of animal experimentation that the results of animal experiments are not transferable to man and are therefore worthless. If one examines the grounds for this statement, one finds examples of cases where, at the start of a new potential treatment, a chance discovery occurs which was not observed in animal research. The discovery of the sulphonamide diuretics is an example. When treating patients with an antibacterial sulphonamide (sulphanilamide) clinicians observed a change in the blood acid-base balance which they explained by the effect of the drug on the kidneys. Thus, a previously unknown effect on the excretion of salts was established, which afforded various new therapeutic possibilities for diuresis. Subsequently, chemically similar compounds were studied in animals. These studies firstly showed that the mechanism of action was based on a previously unknown inhibition of the enzyme carboanhydrase; particularly effective agents were twice developed from this. Further animal experiments revealed that more compounds have an additional effect on diuresis. This group of thiazide diuretics soon replaced all earlier agents which were associated with many side effects, and it became possible to develop short- and long-term diuretics (Pitts, 1959, de Stevens, 1963). This group of drugs is an important part of modern medical therapeutics since they are used in the treatment of oedema, cardiac insufficiency and hypertension. Are animal experiments therefore necessary for some discoveries or not?

From chance to progress
There are a number of examples where unexpected observations, made by an alert creative doctor at the bedside, led to the discovery of new therapies with previously available drugs. Sulphanilamide (discovered by accident in humans) is not itself routinely used for oedema treatment: the action was relatively weak and the necessarily high doses had many undesirable side effects. As discussed above, its diuretic effect was attributed to the inhibition of the carboanhydrase certain enzyme system. Now if such an inhibitory action could be effected in a simple *in vitro* system, i.e. in the test tube, small amounts of blood could be used as the source of enzyme. This *in vitro* method (which could falsely be considered an alternative method) was used to investigate a large number of newly synthesised substances for inhibitory activity, and can be viewed as a test for predicted choice. Parallel animal experiments however, in which urine excretion was measured,

showed occasional marked differences. In some animals, reduction of activity due to metabolic (enzymatic) breakdown of the drug was observed, but in others there was an augmentation of the diuretic effect. Only by combined use of the *in vitro* animal experiments could a suitable compound be selected (acetazolamide). With this agent an effective reduction of oedema was brought about in patients for the first time. Nevertheless, the result was not optimal, since the enzyme inhibition described above led to a 'one-sided' excretion of electrolytes in the urine; sodium was not excreted with the same amount of chloride and therefore diuresis was 'blocked' after a number of administrations. Further animal experiments were needed to obtain a drug which effected the excretion of sodium and chloride through the kidneys in approximately the same amounts.

Transferability good - bad
So far it has proved impossible to characterise the mechanism of action of chlorothiazide in an *in vitro* system. On the other hand, analogous substances can be recognised by a simple and painless animal experiment, the results of which can be extrapolated from rats to man with great accuracy. Transferability does not only extend to the 'excretion pattern' of electrolytes (salts) in urine, but also to the dose of the compound (Lund and Kobinger 1960). In these experiments, the active agent is administered to the animal *via* an oesophageal tube, the urine in the bottom of the cage is collected over six hours and the concentrations of sodium and chloride determined. The animals are replaced in their cages and can be re-used for the next test substance after a week. In this way several substances with an effect similar to that of chlorothiazide were quickly found, but with stronger and longer-acting effects. These substances gave rise to a new treatment: the treatment of hypertension by 'desalination'.

In diuretic research however, poor transferability from animal to man is occasionally noted. The mercury preparations, beset with many side effects, and which were used before the development of sulphonamides, are barely effective on rats. In order to find worthwhile improvements, studies on dogs had to be performed. An important fundamental of drug research arises from this fact: if a new drug is being tested, then it is necessary to study several different animal species before a conclusion can be drawn as to the effects on humans. It would be a mistake - and unscientific - to predict the effect on man from the results of experiments on only one species of animal.

Healthy animals - sick people
There is another argument against animal experimentation in the search for new treatments. Such experiments are undertaken on healthy animals, but drugs are used to fight disease. Indeed there are few diseases which are common to both animals and humans, and research into this field is still only in its infancy (see p. 12).

However, success has been achieved in recent decades in finding effective medicines, for example against heart diseases (angina pectoris) or hypertension and its sequelae (renal and arterial disorders, cardiac and cerebral attacks) as well as analgesic and anti-inflammatory agents, which do not cure sick people or animals, but alleviate their suffering and improve the quality of life. Such compounds were discovered almost exclusively in animal experiments. This was possibly due to:

1) knowledge of the physiological functions, whose dysfunction leads to performance failure, (i.e. to illnesses), and
2) investigation of the chemical agents on these physiological functions.

Thus it was up to the skill of the researcher to analyze the functions into further partial functions, which possess relevance and which constitute an accessible means of measurement.

The beta-adrenoreceptor blockers ('beta-blockers'), are a group of drugs which have markedly improved the treatment of angina pectoris and hypertension. Various animal experiments showed that certain chemical agents (e.g. adrenaline) enhance certain body functions (e.g. the heart rate). This action is produced by stimulation of specific cell constituents, the so-called beta-adrenoreceptors, or 'beta-receptors'. Later, also in animal experiments, compounds were discovered which blocked the beta-receptors, and impeded the action of the stimulatory agent (e.g. adrenaline). The chest pain suffered by patients with angina pectoris is due to a disparity between oxygen utilisation and oxygen supply to the heart muscle. The English pharmacologist, Black, realised that beta-blockers could be used to reduce the heart's demand for oxygen (e.g. by decreasing the heart rate) and began the therapeutic use of beta-blockers (reviewed by Shanks, 1984). Many compounds were tested in animal experiments before the first practical and worthwhile drug - in the form of propranolol - became available to doctors and patients.

The results from numerous animal experiments therefore have to be consolidated - rather like a jigsaw puzzle - to reveal the mechanism of action of a compound. By comparing these data with known substances the effect on humans can be predicted from this battery of results - i.e. deduced from animal experiments. The smaller the divergence from established data, the more certain is the prediction; the newer the data, the less sure it will be. The examples given above show that this method helps sick people.

Is basic research unprofitable?
Animal experiments to find new drugs are accepted by most people; those in basic research are more difficult to come to terms with. It should be noted that knowledge which can be

used as the starting point for the applied research described here is obtained by 'pure' (i.e. not application-orientated) research. The elucidation of functions in organisms and of interrelationships between different organs, first resulted from the desire for knowledge and understanding. All scientific knowledge in medicine has eventually led to some application in the healing arts - and will continue to do so in the future.

Certainly our knowledge about the transferability of results from animals to man cannot deny its research background. The 'beta-receptors' discussed in the preceding section were studied on different species of animals and, since the introduction of the 'beta-blocker', on humans as well. Thus it was established that these cellular components behave identically in every animal species studied, and therefore have an identical molecular structure. The same applies to many other physiological systems and suggests that the most important life-regulating building blocks are the same in both man and animal. These facts enable us to understand the basis upon which we may hope to extrapolate animal experiments to man.

The test pyramid
Hundreds, even thousands of test compounds must be tested before a new drug is found. This is often regarded as purely empirical random testing, associated with the pointless sacrificing of experimental animals.

One concept emerges from each series of experiments. A chance discovery, such as that described on p. 21, may be the basis for this, but it may also be the results of earlier ground work, e.g. a receptor structure, which acts as the starting point. Various molecular models produced by chemists are initially studied on 'simple' test models such as enzyme preparations (see p. 30), isolated organs or easily bred research animals such as mice or rats. The scientist's hypothesis must be formulated clearly, so that compounds which do not meet certain minimum requirements are excluded from further study. Active compounds are further tested for specific effects in more complicated experiments on larger, and therefore higher animals, such as rabbits, cats or dogs. In this way, most of the questions can be answered by observations and measurements on anaesthetized animals. The higher one aspires with a compound on the 'test pyramid' the closer one gets to the ultimate goal, its use on humans. The question of whether an effect previously demonstrated in animal research is comparable with that in man arises at every step. Comparability to compounds tested previously is also an important part of this type of research, and leads to a term often used by opponents of animal experimentation: repeat experiments.

It is necessary to test a drug, the therapeutic value of which is already known, under the same conditions as the new unknown compound. In practice, the conditions are seldom identical to those in which the established drug was first tested. The need for such comparisons is shown by the fact that clinical proving of new drugs is also done in strictly controlled comparative trials with the best currently available drug, in order to detect a potential beneficial effect of the new preparation. Such investigations are now required by the authorities of every developed country before the introduction of a new drug.

Alternative methods - a reorientation
The use of enzyme preparations or organs in the basic 'test pyramid' steps was mentioned above. Amongst these can also be included studies using bacteria, cell tissue culture and other biological tests which are more correctly called complementary methods. They are used expressly in the quest for new drugs, since they can be rapidly performed, and, because of the homogenous biological material, yield highly quantitative data. Transferability to humans is particularly good when the enzyme preparation used is identical to that of man. Sometimes complicated transport processes can be measured directly in human cells. For example, sodium-potassium transport across human red blood cell membranes can be inhibited by cardiac glycosides (compounds resembling digitalis). The intensity of this action makes it possible to predict with certainty the strength of the effect on the human heart (Schatzmann, 1953). It would however, be irresponsible to use a new drug solely on the basis of such a result. Such a test can only be part of the 'jigsaw puzzle', and subsequent investigations on live animals cannot be abandoned. Factors such as the transport of substances from the gastro-intestinal tract to the site of action, its metabolism (breakdown and excretion from kidneys, lungs etc.) and the reciprocal influence of several organ systems (e.g. cardiac, vascular and renal) can only be tested on the intact, entire organism. For this reason the expression 'alternative method' is misleading since such isolated tests only resolve partial aspects of the question. The concept of correlating many such partial aspects by mathematical integration (computers) has been voiced repeatedly by the non-expert. Although this idea is legitimate at a time which has produced great technological advances in many areas, it is however, naive or extremely premature, considering the multiplicity of processes which exist in the smallest biological structures and which remain to be closely studied. Nevertheless, it should be said that 'alternative methods' are being tested and used by researchers. It should also be stressed that for certain stages of drug research, studies on the whole animal cannot be relinquished. However, advances in this field will undoubtedly lead to a reduction in experimentation on the whole animal without any risk to humans.

Cost-benefit ratio: economic aspects
Various criticisms have been directed towards the profit-orientated pharmaceutical industries in recent years. However, unnecessary expenditure was a charge which was seldom laid against them. Studies of test compounds on the whole animal are the most expensive form of drug research, and on financial grounds alone, there is no reason for conducting unnecessary animal experiments. Most experimental animals are bred specially

for research purposes and their rearing is demanding and expensive. Table 1 gives a view of the costs, and illustrates three things:

1. Animal experiments should generally be reduced to a minimum.
2. In the outlay given, the 'lower' (cheaper) research animals are preferred to the 'higher'. This principle has already been mentioned in the section 'the test pyramid'.
3. If possible, one of the considerably cheaper alternative or complementary methods which were discussed in the section 'alternative methods - a reorientation' should be used.

Table 1. Costs of the animal specially bred for research purposes (the position in 1985):

	DM		
One mouse	2.50	-	5.00
One rat	6.00	-	20.00
One rabbit	37.00	-	100.00
One cat	500.00	-	750.00
One beagle	800.00	-	1600.00
One monkey (rhesus)	7500.00	-	11250.00

A long-term cancer study for one substance over two and a half years costs 1.5 million DM.

The number of animals used in drug research in recent years has declined markedly. For the drug manufacturers in the Federal German Republic, following reviews of the Federal Association of the Pharmaceutical Industry, the percentage reduction in animal usage has recently been calculated (Table 2). How can this sharp reduction be explained?

1. The improvement in breeding and maintaining research animals makes homogenous groups of animals available to the researcher. This narrows the distribution of results enabling statistical significance to be achieved with a smaller number of animals.

2. The additional inclusion of alternative methods (see section 'Alternative methods - a reorientation'). Such methods are already being used by drug researchers, together with the advantages (homogenous material, quantitative results and rapid performance) discussed above. They are considerably cheaper and quicker than experiments on the whole animal.

Table 2. Percentage decline in the use of research animals in the pharmaceutical industry (Democratic German Republic). The values for 1977 were arbitrarily set at 100 in each row for comparison.

	1977	1978	1979	1980	1981	1982	1983	1984
Mice	100	90	83	80	68	66	65	58
Rats	100	94	86	71	63	64	63	62
Guinea-pigs	100	88	81	56	55	56	55	55
Hamsters	100	74	69	43	42	40	36	24
Rabbits	100	87	79	48	44	45	42	41
Cats	100	84	72	52	47	40	37	29
Beagles	100	92	81					
Mongrel dogs	100	94	82	67	57	58	57	52
Other dogs	100	114	99					
Pigs	100	206		221	264	262	290	288
Sheep	100	87	100	under 'other animals' from				
Goats	100	62	109	1980 (see below)				
Cattle	100	93	102	79	80	82	104	89
Horses	100	80	128	0	0	10	10	2
Monkeys	100	129	141	46	37	43	45	46
Other mammals	100	39	37					
Birds	100	120	133	Other animals (from 1980)				
				93	99	87	84	91
Cold-blooded animals	100	107	121					
All animals	100	92	85	75	66	64	63	59

Of particular note in Table 2 is the increased use of pigs. This may be attributable to the problem-free nature of this species and their availability. Alternatively, one may recognise a concession to the emotions of those who see cats and dogs as lovable pets, but pigs as meat-producing and relatively unattractive animals. In any case, these figures should give rise to reflection; why should laboratory rats be considered worth protecting, while their wild counterparts are destroyed by different, but often much more gruesome methods (Staberl, 1986)?

New drugs, as well as other therapies, which can only be produced using animal experiments, also have a politico-economic aspect. Thus, a shorter hospital stay, whether it be due to improved surgical techniques or to pharmacotherapy, is an economically important factor which affects the general public. The frequently complained of multitude of drugs should be seen in this light: analogous compounds do not always represent a medical improvement, although they can, because of greater efficacy (= smaller doses) or cheaper production, bring economic relief for patients and the taxpayer. Such compounds also require basic studies in animals, and an economic consideration of the public's viewpoint is to be recommended to critics of drug proliferation.

References

Lund, F.J. and Kobinger, W. Acta Pharmacol. (Kbh.) 16 (1960) 297.

Pitts, R.F. The physiological basis of diuretic therapy. Springfield, Illinois 1959.

Schatzmann, H.J. Helv. Physiol. Pharmacol. Acta 11 (1953) 346.

Shanks, R.G. The discovery of beta adrenoceptor blocking drugs. In: Discoveries in Pharmacol. M.J. Parnham, J. Bruiwels: vol. II. Elsevier, Amsterdam 1984.

Staberl. Neue Kronen Ztg. Nr 9395 (1986) 6.

De Stevens, G. Diuretics, Chemistry and Pharmacology. Academic Press, New York 1963.

Tables 1 and 2 from: "Tiere in der arzneimittelforschung. Nutzen und grenzen von tierversuchen und anderen experimentellen modellen." In: Bundesverband der Pharmazeutischen Industrie e.V., Frankfurt, 3. überarb. Aufl. 1986.

3

Toxicology: Safety for Man and Animal

In 1538, Paracelsus first showed that the concept of 'poison' is closely related to that of 'dosage'. Even substances which are essential for human life, can develop toxic activity if taken in excess (Figure 2).

Figure 2. Pharmacologists and toxicologists mainly use rats. Pharmacologists (a) handle these animals to study their behaviour or to record various functions under anaesthetic. Toxicologists (b) are primarily interested in the effect of the poison which they wish to investigate and use many ingenious methods to achieve this in rats or in one of the organs removed from the rat. (R. Dahl: *Der krumme Hund*. Rowohlt Taschenbuch 959).

This applies for example, to table salt, iron, vitamins and hormones, and not just to medicines and substances encountered in our environment. In a chemical world, with all its associated benefits, protection against injury or damage to our health is only possible through knowledge of the potential dangers, and of the different kinds of chemicals which can induce them. This assurance can only be guaranteed by appropriate toxicological experiments in animals. However, it may also be achieved by using the more sophisticated alternative methods of modern toxicology, rather than through the deaths of many animals.

Dependence on animal research
(Drug safety)
R. Czok

The doctor treating a condition and handling a new drug, has both cause and obligation to consider its advantage in the course of treatment. This is inextricably associated with consideration of the safety of medicines. Even those drugs which have been used for decades do not necessarily help all patients at any given time. There are exceptions, where the desired effect either does not occur, or where an unwanted side effect is particularly severe.

Treatment with a new drug or with a compound currently under pharmaceutical development is only justified on medical grounds if the anticipated advantage to the patient is greater than any possible disadvantages. Ethical foundations and legal provisions are required. In many countries, clinical trials of the medicine must be carried out before its administration to healthy or sick people. Together with the results of earlier investigations, the realization of clinical trials must be based on the latest state of scientific knowledge and individual experience. Neglect of these fundamental principles is a punishable offence.

In most countries, new drugs which have not been approved for commercial use, must be sanctioned by an ethical committee before they can be used by doctors on healthy or sick subjects. The results of preliminary studies are presented to the committee, and doctors, pharmacologists, toxicologists and sometimes lawyers, have to satisfy themselves, that as far as can be judged, the patient being treated will benefit, and the healthy volunteer or test person will suffer no detriment to health.

Safety considerations for patients or volunteers assume particular importance when a drug is to be used for the first time. Side effects, whether previously observed as an enzymatic reaction in cell-free systems, in single cells or confirmed in laboratory animal experiments, do not always not occur in humans in the same way, to the same extent and with the same duration.

Benefits to the patient are expected from the administration of a drug; disadvantages could arise from the properties of the drug and from its erroneous use. Considerations of the safer use of a drug have therefore two aspects of note: efficacy and tolerance.

The therapeutic effect
A patient needs treatment. A doctor may only institute a new type of treatment if an equivalent or superior healing or alleviation is anticipated. This principle applies in every situation, and is particularly important if a future drug should be used for which no predictable results have yet been elicited from clinical investigations. Before the doctor makes up his mind, the manufacturer of a potential drug must prove whether the known facts about the agent and its effects are sufficient to justify its clinical use.

This evaluation of a compound can only be supported by preclinical investigations on models. In many such instances, a new drug is often chosen for testing on simple models without using living creatures. Cell-free systems of this kind may be used successfully when very precise knowledge of the mechanism of an effect already exists. It is usually the result of numerous previous animal experiments. Lower life forms such as viruses, bacteria or parasites can act as reaction models, for example in chemotherapy. Such models have the advantage of demonstrating, in a technically simple way, the type of action and the characteristics of a compound. However, they also have the disadvantage of over-simplification. The therapeutic action of a substance is not merely dependent on the type of effect it has at the site of application, but also on whether it remains chemically unchanged, in a sufficiently large quantity or concentration, over a sufficient length of time at the site. The degree of resorption and distribution in the patient, and its length of durability as an active agent (pharmacokinetics) determine its therapeutic success. The functions of the gastrointestinal tract as an entry point, the blood circulation system in distribution and the internal organs in the metabolism of the drug (i.e. stomach, liver and lungs) all work together. For each of these functional components there is a model amongst the so-called alternative procedures. However, no model is as good as the laboratory animal in simulating the complexity of the combined working effect in man.

Before the doctor can administer a trial drug therapeutically, it must have, for example, demonstrated its chemotherapeutic action in the infected laboratory animal, lowered blood sugar in the diabetic animal, or reduced experimental hypertension if it is later to help a patient in these respects. At the very least it must be equal to drugs already in use, and should additionally hold out the promise of better results. The manufacturer evaluates the latter point very critically; a compound which offers no therapeutic advantage over drugs already in use ('me-too-product') will probably not be a good saleable product, and the costs of testing and development will be high (c. 120 million DM).

Nine out of ten compounds fail in the effect-proving phase in the animal model, and do not get as far as a clinical trial. This increases safety, even if, amongst the discarded drugs there might have been some which would have been of particular value to patients. However, the inadequacy of the model has not allowed this special property to be recognised.

How infallible is knowledge obtained from animal studies relating to human therapy? The small laboratory animal (mouse, rat, guinea-pig, rabbit) differs from man. Body surface to body weight ratio is greater, heart rate is higher, metabolic rate is more rapid and life-span is shorter. In the animal however, these and other differences also alter the action of known drugs whose action in man is very well known. These drugs may therefore be used as standards. Comparison of the test compound with a standard drug in the animal model permits reliable conclusions to be made about its effects on patients. The less the test compound varies chemically from the standard, the safer these predictions become.

The best model for the patient is undoubtedly the healthy volunteer. With his cooperation, knowledge is gained about the behaviour of the compound in man (resorption, pharmacokinetics and biotransformation), and pharmacological effects are measurable which are of crucial significance for the treatment of the patient. This knowledge considerably reinforces the extrapolation of the animal model to the patient and increases the safety of the drug.

Safety

Since the time of Paracelsus, his saying "*All things are poison and nothing is without poison. The dose alone renders a thing without poison*", has been proven many times and in some alarming ways. This applies not only to drugs and beverages, the intake of which presupposes intent, but above all to environmental factors and foodstuffs which we cannot avoid. The decision to take a medicine must take into account centuries of experience gained from toxicological observations and experiments which form the totality of the present state of scientific knowledge. Their use requires ethical bases and legal prerequisites.

The general toxicity of substances and their action as poisons can be investigated on pure material in the laboratory animal. Cell-free models such as enzymes, enzyme complexes, or cellular organelles such as mitochondria and microsomes, as well as cell cultures and isolated organs, may be used successfully in tests for a *specific* kind of toxicity (see also p. 57). Their relevance for adverse effects in man must firstly have been deduced from clinical observations and experiments on animals. For example, penicillin would be a convulsive poison like strychnine if it could impinge on the brain. It is only non-toxic because it cannot get past the blood/brain barrier. From this it is clear that a toxicity test on isolated brain tissue would lead to a completely false result. These models are most useful with the least known compounds chosen from a group with similar action. With their help, numerous animal experiments may be avoided. Nevertheless, both therapeutic action and undesirable side effects are of crucial significance, as well as the kind of incompatibility and degree of toxicity, and factors such as resorption, distribution, metabolism and excretion.

The possibility of fertility disorders, embryo toxicity, teratogenicity and congenital disorders due to a test drug can only be reliably investigated by experiments on laboratory animals (see also p. 166). The formation and development of foetal organs at the correct time, and their influence on each other, presuppose a hormonal regulation, which although recognised, is not understood in detail and thus cannot be simulated in the model. Cells or embryos from the animal can be kept in nutrient broth for a limited time and will develop. Thus, the experimental effect of the test agent on growth and maturation may be assessed, and long-term growth, morphology of organs and survival compared with the untreated embryo (as a control) can be evaluated. A positive result in experiments on explanted rat embryos has considerable significance and protects both laboratory animals and man from further extensive trials with this compound.

All experiments on the laboratory animal are aimed at ascertaining a 'no toxic effect level' (NTEL); i.e. that dose which, under given conditions, manifests no toxic effect on behaviour, growth and organ functions, so that the animal being given the test drug remains quite healthy.

NTEL investigations are performed on at least two species of animal, and the results from the most sensitive are taken as limiting. The length of the animal experiment is dictated by the duration of therapeutic use. If it is long (e.g. contraceptive agents, or drugs for treating diabetes or high blood pressure), the trial on the small laboratory animal covers its entire life span, and may therefore reveal sensitivities in the embryonic, infant, adolescent and adult phases of life.

A particular problem of drug safety relates to side effects which first become clear very late in the treatment, or after it is concluded. This group of problems includes tumorigenesis, which, from the first phase of initiation, through promotion and progression, often takes decades to develop in man. The same mechanisms and phases of development also take place in the small laboratory animal (e.g. mouse and rat), but within its shorter life span of 2-3 years. They are the one suitable model for man as they are small enough to be examined in large numbers, and the result occurs early enough before the clinical use of a new drug. Furthermore, they can and must be given dosages which lie within the toxic range, and which are therefore much higher than they will subsequently be for the patient. At least two species must be investigated. If no increase of naturally occurring tumours is seen in either species, the danger of tumorigenesis in man may be considered as very small.

As tumours generally only have a very low incidence (less than 10% in rats), testing for carcinogenesis is protracted, expensive, and requires many animals. For decades therefore, strenuous efforts have been made to develop alternative methods. For example, *in vitro* experiments on microorganisms (Ames test), on cell cultures (transformation test), on surviving cells in culture (chromosome aberration, sister-chromosome exchange) may indicate the hazard of a test compound. Nevertheless, numerous experiments with these models still do not give the degree of certainty achieved with animal experiments; a legal requirement in specific circumstances.

Toxicological investigations on several species of animal not only aim to identify dosages likely to be harmless in man, but also to determine the type of toxicity which can lead to side effects, either at too high a dose or under other unfavourable conditions. The most sensitive organ and the most easily disturbed metabolic pathways may be identified using toxic doses in laboratory animals. Special attention is then paid to these in subsequent clinical investigation.

Two types of reactions occur between compounds and the organism to which they are introduced:

1. The drug alters the function of an organ in a way which is therapeutically significant in a specific dose, but toxic in a higher dose.

2. The organism alters the substance chemically, whereby it becomes ineffective and can be excreted through the kidneys. Alternatively, the substance may be metabolised in such a manner that, from an inactive parent compound, a pharmacologically active substance is formed which is responsible for toxic effects. Certainly these changes, which lead to the detoxification or 'toxification' of a drug, must be determined carefully during the drug's development. Many such chemical reactions and toxicologically-questionable chemical structures are known, and are now too numerous to be easily remembered. Thus, computers may be used to store and retrieve this knowledge, and, by rearranging known data, demonstrate new kinds of reactions which are likely to occur in the animal. Thus, the extent of laboratory animal investigations may be considerably reduced.

Biotransformation and excretion, together with resorption and distribution determine the behaviour (pharmacokinetics) of the compound in the organism in terms of time, amount and concentration. With animal research, the characteristic parameters for the definition of kinetic reactions as they concern toxicity may be ascertained. It is an unconditional prerequisite that these should be known before a drug is first used on humans. The safe dosage and the frequency of permitted administration reflect the value of these experiments. However, these considerations are only applicable if the pharmacokinetics of the test drug in humans is identical with that already determined in the laboratory animal.

When the compound is first given to man, healthy subjects are used to establish the extent to which the results of animal experiments can be extrapolated to humans. This involves determining the pharmacokinetic and pharmacodynamic parameters as well as biotransformation. After careful evaluation of the clinical observations, and clinico-chemical analyses of the pharmacokinetic data, it may be decided whether a higher dose may be given and over what length of time. Here again, the knowledge gained from animal experiments with longer-term administration will be used.

The first use of a new drug in humans always carries the risk of side effects. On both ethical and legal grounds, this risk must be kept as low as possible. This requires comprehensive knowledge of the substance and its behaviour in the mammal, which can only be obtained from appropriate mammalian investigations.

Toxicity of lipsticks?
Safety of skin care products, food additives and agents in daily use
F. Lembeck

Lipsticks are particularly ideal test systems for chronic toxicity in humans. The dyestuffs that they contain are presented daily in high concentrations to the sensitive skin of the lips. The composition of the lipstick guarantees prolonged contact, and small amounts of the dye are inevitably swallowed. Their use remains uninterrupted during pregnancy and illness. In the whole field of animal experimentation, no other such extreme example of chronic application is known.

The use of lipsticks has only been made possible because their dyes and other constituents undergo thorough toxicological testing over a long period of time. Is it still ethically permissible for so many experimental animals to be sacrificed for these luxury articles?

The lipstick is a decorative cosmetic. Unless he is an actor, a man can refrain from using this decoration. In general however, he enjoys the company of a lady whose health and physical wellbeing, and consequent good humour, is accentuated by lipstick.

It may be added however that the lipstick serves not only as decoration but also as protection for the lips, as some lipsticks have protective qualities against sunlight. The author finds lipstick and other items of make-up offensive only when the lady using them is accompanied by a dog from which the most rudimentary hygiene, and food appropriate to its species are withheld.

Cosmetics are used for the cleaning and care of the skin, hair and teeth, and for keeping them healthy. In our modern world, in which many people have to live in close proximity to each other for many hours of the day - at work or on public transport - general hygiene and body care do not only concern the individual. One can imagine the daily ride in the suburban train in the last century: shabby, dirty people, covered with scratch marks, with clothes, head and pubic lice, scurfy-faced (impetigo contagiosa), unshaven and smelling of drink! *"One not only sees poverty, one actually smells it"* said one of my clinical teachers. It should not be forgotten that only 40 years ago, the dusting of passengers with DDT was authorised at certain frontiers. Great success was achieved with this measure, because by destroying lice an epidemic of typhus in Naples was halted.

The puritanical attitude is that reasonable body care can be achieved with plenty of water, good quality soap, oil for skin care and powder for cleaning the teeth. However, the individual feels better if he or she uses a soap which really suits the skin, a skin cream which is appropriate to the physiological features of the skin and its particular needs (housekeeping, nursing, technical work), if unacceptable body odour is prevented by a deodorant, if the hair shampoo corresponds to the drier or greasier scalp and if the toothpaste really cleans and deals with tartar without damaging the dental enamel.

Such considerations make it clear that reasonable modern body care cannot be achieved without using a fairly large number of agents. Thus, it becomes essential that such agents are banned from use in cosmetics agents if they carry an unacceptable health risk.

How can this be achieved without animal experimentation? The testing is difficult enough, since all these substances are used long-term, often on large areas of skin surface, by small children and sick people or by those with a tendency to allergies. The last possibility has to be considered since the development of allergies occurs much more readily in a hairdresser for example, since she, unlike her customer who comes every three months, comes into daily contact with hair dyes.

There are now national and international guidelines for the toxicological and dermatological testing of cosmetic agents. As a first priority all substances in a cosmetic are tested toxicologically, and the finished product is tested for accordance with certain prerequisites (for amplification see Grunow et al 1986). The leading manufacturers of cosmetics are painfully intent on producing irreproachable products, since a very few untoward incidents would persuade the customer to use one of the many competitors' products.

Testing the components of a cosmetic on animals is unavoidable because possible serious hazards - carcinogenicity or liver toxicity - are not easily recognised even after years of application in humans, especially since such effects are not necessarily visible externally. Modern animal experimental investigations are adjusted to the use to which the product is to be put. Chronic effects are assessed by giving small laboratory animals (rats, mice, hamsters, guinea-pigs) the test compound in food and drinking water. Their body weight is recorded, the blood count examined and endogenous substances in blood and urine analyzed, from which the effect on internal organs (e.g. liver or kidneys) can be assessed. Other tests embrace carcinogenicity, mutagenicity and allergic components. Most tests are not stressful for the experimental animal, since it is not toxicity which is primarily tested, but the dosage which will cause the first signs of any possible damage. At the end of these long-term investigations the animal is killed under anaesthetic for histological examination of the organs.

Certain animal protection societies briefly distributed 'white lists' of those firms which do not use animal experiments in the production or sale of their cosmetics. Only the insider knows that this is not due to skill, since the acute toxicity (LD_{50}, see p. 100) of the components has previously been assessed through animal experiments, and can be found at any time in the Handbook of Toxicology! In addition, the comment of some manufacturers, that only pure herbal dyes, perfumes and active substances are contained in their cosmetic preparations may reassure the customer. However, the toxicologist sees this in another way, since he knows how many toxic substances are obtained from plants, the effects of which can be judged not with our senses, but only by biochemical or pharmacological methods.

Our skin provides a great deal of protection against heat, cold and mechanical insults, but this is not true for the delicate mucous membranes around the eyes, nose, lips and genitals. These mucous membranes are densely innervated and react with pain, reddening and inflammation to slight irritation. As anyone who has bathed in the ocean can attest, a 3% solution of table salt burns the eyes very strongly. A test based on the irritation of the mucous membrane of rabbits' eyes, which are particularly sensitive (Draize test, see

p. 102), has recently led to considerable agitation about animal torture. Alternative methods for this test have been, and are still being, sought, but their reliability must of course be tested first. In the author's opinion it is not primarily a question of whether the approved Draize test or an alternative method is used, but of how this test is performed. If a test compound is used in a very low concentration, and this concentration is increased only up to the appearance of symptoms of irritation, but not as far as to cause tissue damage, then the necessary information has been obtained; i.e. the limit at which the compound is perfectly tolerable to the mucous membrane. Using such a procedure, rabbits' eyes are irritated less than those of a non-smoker in a pub where he is subjected to cigarette smoke irritation for hours.

The efficacy and safety of cosmetics is often the subject of very complicated investigations by dermatologists, dentists or allergy specialists. These investigations serve not only in the control of the effect, but also for the further development of a preparation. This development serves not only to improve the compound but also in the introduction of new ingredients.

Reference
Grunow, W., Schleusener, A. and Vivell, K. Notwendigkeit von tierversuchen zur gesundheitlichen beurteilung von kosmetischen mitteln. Bundesgesundheitsbl. 29 (1986) 1-6.

Poisons by the ton
Safety of industrial chemicals
D. Henschler

The professor sends his laboratory assistant into the washroom and closes the door. He tells him to trickle a fluid onto the floor and to fan it with newspaper long enough for it to evaporate. He then sucks the atmosphere created in the 'test room' through a flask using a manually-operated air pump. The volumes being sucked through are measured, enabling the subsequent concentration of the test substance to be determined, i.e. after chemical analysis of the flask contents. Periodically, the professor looks through the window at the condition of his test subject. If he leaves the washroom with clear signs of survival after 30-60 minutes, and is fit for work the next morning, then the test concentration is considered to be 'maximally' acceptable in the work place.

Safety limits as a protection principle
Using such experiments in 1886, the Bavarian Karl Bernhard Lehmann - Professor of Hygiene at Wurzburg University - initiated efforts towards the protection of industrial

workers from damage to their health through chemical agents at work. It was science and industry's answer to Bismarck's social legislation in the Kaiser's Germany, which obliged the employer to compensate occupational illness through state-regulated insurance arrangements. The principle was simple: subjective signs of danger or objective symptoms of disease due to industrial poisons should set the limits - as dosage or concentration - which adequately guarantee the protection and preservation of the ability to work. Thus, the safety limit principle became a legislative tool, worked out and based on toxicology as a rising science. Man was the first and, at the same time, the most reliable test animal. Thus, he could appreciate and report the actions of the test compound on the mucous membranes of the eyes and nose, and on respiration etc., better than any other creature. Even in those days of the first industrial revolution, health concerns were centred on medical observation and investigation into dangerous work places in the chemical industry. Initially, the dye manufacturers, and later the organic chemical industry, set up clinics to deal with acute poisonings, but in others, safety measures and far-reaching provisions for care would be developed by bringing together such observations.

New exigencies in work protection: the pressure of innovation
However, this experiment soon proved to be inadequate in helping people to avoid health risks. Karl Bernhard Lehmann, who together with Ferdinand Flury had over four decades established safety limits for many compounds in the workplace, began experiments on research animals. Two reasons are generally given for the fact that animal experiments now dominate toxicological risk assessments in the workplace, but also by extension, many other areas associated with chemicals:

1. From the middle of the last century to the middle of this century methodological developments in chemistry have proceeded constantly. Since then however, progress has accelerated at a previously unimaginable rate. This is due to two main factors. Firstly, there are the innovations, i.e. the chemicals newly introduced into production and marketing. This is based on chemical research which is mainly directed at the synthesis of new compounds. Details of this growth tendency are given by the Registers of Chemical Abstracts, which list all new compounds in each separate country, and are cited in the chemical literature and in the Patents Notifications. This development over the last third of this century is shown in Figure 3. The basis for this enormous acceleration during the last 10 years alone is the rationalisation and optimisation of research methods, especially in industrial laboratories, which today operate largely by computer-controlled syntheses. The presence of competition in this production-intensive industry leads one to expect a continuation of this growth effect for the foreseeable future. Chemical investigations extend not only to new compounds with new areas of use however, but also to improved production, more effective use and enhanced chemical

reactivity of new compounds. Every chemist who has been actively researching for twenty years or more is familiar with this trend merely by comparing the laboratory chemicals introduced for chemical analysis then and today: derivation reagents for gas chromatographic analyses display reaction potentials unimaginable to our forebears. In general, the damage potential for biological systems, and thus for man, also rises with chemical reactivity. The constantly growing number and reaction potential of modern chemicals are one of the biggest toxicological challenges in developing testing strategies to reliably evaluate the danger to man, before he comes into contact with the substance - in whatever form.

The number of chemical innovations do not necessarily signify a proportionately increased risk for humans. How many of those new compounds entered in the chemical registers are actually in practical use is not reliably known. Previously, new laws prescribed the notification and authorization, and thus the systematic acceptance by the authorities. However, the same exponential increase is also plainly seen in chemical production. Figure 3 shows figures for the Federal Republic of Germany. Thus, the tonnages, which have run into millions since the end of the 1960s, are still rising much more sharply than the number of new compounds. This indicates an expansion of production volumes for the average individual product.

Does this imply the same health risk for people in highly developed countries? This question is not going to be resolved using the available data, and measurements of individual and collective exposure are required. A direct comparison with production figures and human exposure is most readily possible by looking at drug usage (see Figure 3). Astonishingly, the growth rates in this sector, measured by volume of sales, are clearly even higher than all organic chemicals. Furthermore, with drugs there is also a perceptible tendency, based on further scientific development and the requirements of the medical profession, for increased drug potency. This may involve both the desired therapeutic effect and the undesirable side effects.

One may postulate different attitudes to these figures and development tendencies. There are certain arguments for limiting the previously quasi-autochthonously inspired growth of innovation and production. The benefit-risk assessment for all chemicals, called for equally by science, politics and government today, and which is performed for diverse usages on quite different points of view, repeatedly points to superabundance and abandonment. In the final analysis however, the usage and growth rates in liberal economic systems depend on consumer habits and demand. As a science, toxicology must face up to the challenges and must examine and evaluate the safety of chemicals. The figures alone invite reconsideration of conventional testing strategies. Progress can only be achieved by quicker and less costly methods, but this requires further animal experiments.

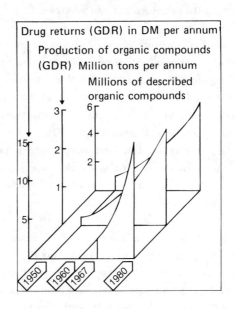

Figure 3. The tendency towards new organic and technical chemicals, shown by the increase in 'Chemical Abstracts' of reported chemical compounds (rear profiles), of the production of organic chemicals (mid profiles) and of drug returns according to the Federal Association of the Pharmaceutical Industry in the German Democratic Republic (fore profiles).

2. Previously, toxicology was mainly concerned with the analysis, therapy and protection against acute poisonings. These are characterized by rapid onset, a well-defined episodic association and by symptoms which are intense enough for diagnosis and which are typical for certain chemicals or a group of chemicals. With increasing numbers of chemicals in the environment however, particularly those with high persistence, chronic poisonings are increasingly of major concern to man, and are now the main occupation of toxicologists in industry, government and universities. Chronic toxicity is difficult to evaluate since, because the symptoms of disease first appear long after exposure to small concentrations of the toxin over a long period of time, the association is no longer clear. In addition they are generally not very characteristic and are often difficult to distinguish from everyday ailments.

New tyoe of action: genotoxicity
A special kind of toxicity in particular has come to the attention of scientists, the public and politicians relatively recently. In specialist terms, this is known as "genotoxic effects" (see also pp. 130 and 165). It concerns cancer-causing and genetic damage-causing effects of chemicals (carcinogenicity and mutagenicity). They are distinguished from conventional injurious effects by several criteria:

- There is a long period of time between exposure and the onset of disease (cancer or hereditary disease). This may be 30 to 40 years for lung cancer in asbestos workers, or after several generations for some genetic diseases. This makes the recognition of causative associations extremely difficult.

- Genotoxic effects also persist after the smallest doses and can be incremented by subsequent effects. From this it may be deduced that - *inter alia* - effects are irreversible, and that firm threshold limits cannot be fixed.

- The embryo can also be damaged by such substances; from all the knowledge obtained so far it actually seems to be more sensitive than the full-grown organism, because it is not yet fully capable of primary damage repair.

- Today cancer is still more difficult to cure than most other diseases, and for the victim this - in combination with the tragic course of the disease in its terminal stages - is a particularly heavy burden.

Maximum Workplace Concentrations: for and against
For conventional, i.e. non-genotoxic toxic effects, science has introduced the principle of threshold values. Protection at work was a leading element here and this laid down the ground rules. Continuous threshold concentrations in the workplace in an eight hour day are termed Maximum Workplace Concentrations (MWC values). Originally a German device (see above), they are now used world-wide as the actual instrument of protection against occupational diseases due to chemicals. They are not 'certain' in lay public opinion, since they can only include hypersensitivity as a result of a reaction tendency in some individuals inasmuch as appropriate observations are reported. Very rare cases of hypersensitivity are often discovered very late, and not necessarily by the doctor. Modern diagnostic procedures which increasingly rely on biochemical parameters, and which show deviations from normality well in advance of individual injurious effects - i.e. before significant disease occurs - have contributed greatly to the current lowering of MWC values. Health protection has thus naturally improved. Not least, currently increasing

health awareness requires scrutiny of the concept of disease and a firming up of the boundaries between health and acceptable levels of discomfort.

Maximum workplace concentrations are today almost exclusively deduced from animal experiments. Human studies to determine MWC values become less useful with increasing use of protective measures and with decreasing exposure to these substances. MWC values are quantitative parameters. They are laid down in steps from factor 2 to factor 2.5. This requires precise dose-effect relationships as a basis. There are at present no *in vitro* methods which provide the required precision of limit setting. The only way to obtain the required data for human protection are experiments with intact animals, the duration of which in terms of exposure and follow-up time depends on the situation, but can vary from a few months to the whole life span of the animal. Renouncing animal experiments would mean workers running unknown risks, as well as the complete abandonment of the limit values principle, which is indeed the best and most effective preventive measure.

Safety limits for carcinogenic agents?
Toxicology can presently formulate no limit values for genotoxic substances. Thus, industrial chemicals have merely to be designated as carcinogenic and contact with them must be kept as low as possible or avoided altogether. In the Federal Republic of Germany, and in some other countries, another kind of limit value has been introduced, since it is required by the practice of protection at work: the TRC (technical recommended concentration) value. In establishing these values, socio-economic, technological and analytical factors are particularly considered. In the end they are a socio-political compromise, which contains a largely unquantifiable residual risk. This is one solution which presupposes another, rather political type of risk-benefit assessment and which satisfies many, although it is not convincing from the scientific or medical ethics standpoint. All the same, it was thought possible to live with it satisfactorily, since the possibility of the checking and lowering of industrial health protection values continually allows for improvement.

The dilemma: lack of quantitation?
In the field of genotoxic substances and effects, developments 10-15 years ago paved the way for the use of substitute *in vitro* tests. The use of microbial mutagenesis testing and cell transformation tests has greatly reduced the numbers of laboratory animals in carcinogenesis and mutagenesis testing (see p. 83). Increasing work in this field has resulted in rapid collection of data on the genotoxicity of increasing numbers of substances. The result is a development still unforeseeable 10-20 years ago; year by year, increasing numbers of substances are classified as carcinogenic and mutagenic. At present, chemicals in the German MWC value lists designated as definite, or strongly suspected carcinogens, comprise more than one third of all listed chemicals. It is conceivable that

soon more than half, perhaps the majority of compounds will be considered genotoxic. This will make all regulatory mechanisms in the future unreliable; unless one is prepared to forego chemicals in any form.

The establishment of new formulae for legislatively solving these problems is the most pressing current toxicological preoccupation. It involves quantifying carcinogenic and mutagenic risks and, with the help of quantitative data, setting new priorities in both protection against chemicals and in chemical developments in general. How are quantitative risk data to be obtained? The most reliable methods are epidemiological studies. If they are carried out on a quantitative basis, i.e. if exposure data are correlated with the incidence of cancer in that population and the corresponding dose-effect curves calculated, the probability can be established of developing cancer in relation to particular levels of exposure. However, there are narrow limits to this approach. Thus, most of these quantitative exposure data are not available. The precision and reliability of the evidence also depends on the size of the population available for study. Often it is too small, and reliable data cannot therefore be obtained. In reality, reliable quantitative data presently exist for only few carcinogens; inhaled cigarette smoke (lung cancer), aflatoxin intake (primary liver cancer), asbestos (mesothelioma and lung cancer) and some cytotoxic agents used in cancer therapy. Add to this the fact that comprehensive epidemiological data can only first be obtained if damage has already occurred, and that protective measures for the first victims cannot then be provided. For this reason alternative strategies must be sought urgently.

As a practicable alternative method to epidemiology, studies in intact laboratory animals have already been considered. All other 'alternative methods' are not able to offer quantitative risk data. Without animal experiments we would have no comparable human risk data for many compounds, and their future development affords a supplementary resource for pharmacokinetic and metabolic studies. These must be combined with *in vitro* studies to highlight the differences between experimental animals and man. Only in this way can the potential for extrapolating data from research animals to man be quantitatively improved. A full scale renunciation of animal experimentation is not foreseen, although there will be a distinct reduction, which will be determined by progress in the use of *in vitro* methods.

Safety belts and crash barriers
New methods in toxicology: cell culture
H. Marquardt

Aims of toxicology
Toxicology is the study of toxins and toxic effects; the concept of toxicity is associated with both the nature of chemical compounds and with their dosages. The latter is often forgotten at a time when picogram quantities (10^{-12} gram) or less of a substance are readily measurable. As an applied science, toxicology is accordingly concerned with (1) the chemical-biological interactions with acute or chronic ill effects on health, (2) the quantitative determination of the health-damaging potential of chemicals, (3) the identification of toxic symptoms, (4) the treatment of toxic effects and (5) protection against toxic agents. Thus, central to the aims of toxicology are the safety assessments of drugs, skin care products, food additives, household products, agricultural and industrial chemicals etc. To this extent, toxicology is of utmost - and socio-political - importance since the nation's health is affected. It may be argued whether cosmetics, detergents and food additives etc. are actually superfluous, although such a debate is irrelevant since the products are part of our life, and will remain so for the foreseeable future. It is the toxicologist's duty to ensure that the user faces no undue risk from exposure to the substances in these products. This is increasingly necessary because the stress on man and his environment has increased considerably in the last 100 years.

To determine potential chemical health risks for humans, tests are carried out for:
- acute toxicity (symptoms after a single administration),
- subacute toxicity (repeated doses over several days or weeks),
- chronic toxicity (doses over all or most of the life-span),
- special toxicity (skin tolerance, eye inflammation, fertility disorders and embryotoxic effects etc.).

From such studies, 'no effect' dosages can be calculated, from which, with attention paid to safety factors, 'acceptable' concentrations for human exposure may be calculated. This general toxicological evaluation relies on the assumption that the person, as well as the most sensitive animal species, react to chemical substances. It is, however, associated with certain limitations:

1. The 'ideal' choice of the most sensitive animal species, can seldom be fully attained.

2. Laboratory animals are different to humans, not just in body size, but also in metabolism, biotransformation and pharmacokinetics.

3. There are no generally binding functional or morphological criteria for the establishment of 'no effect' concentrations.

4. The quantitative or qualitative limitations of animal experiments (e.g. limited number of animals, suitable animal models) must be considered.

These limitations mean that the extrapolation of animal experiment results to man requires careful expert interpretation. It must be stressed that toxicology in general, and the spectrum of toxicological methods in particular, are subject to constant change. The first hundred years of toxicological research was overburdened with *acute* poisonings (syndromes which are characterised by typical symptoms and clearly identifiable connections between toxin intake and onset of disease). Today however, the central interest is *chronic* poisoning (ingestion of the toxic agent, often in trace amounts, over longer periods of time). The evaluation of these toxicities is a disproportionately more difficult problem, which is necessarily associated with the need for new methodologies. The development and introduction of new testing strategies with the inclusion of appropriate short term tests (also misleadingly termed 'alternative methods' in connection with the animal protection movement) and mechanical studies, is inescapable.

'Alternative methods'
Alternative experimental methods must include non-biological (mathematical models of structure-activity relationships, computer models of biological effects etc.) as well as biological methods. However, in the following discussion, only the biological methods (*in vitro* cultures of bacteria, isolated mammalian cells and whole organs) will be considered. These cell culture methods have generally been developed in biomedical research to answer scientific questions more efficiently. The detailed data from studies on the cellular level of metabolism helps to explain the mechanisms of toxic effects on the molecular level, which is impossible on the complete animal (man) because of numerous interfering factors (e.g. the presence of the endocrine and immune systems etc.). On the other hand, this highlights the inherent weakness of these methods; they only represent part of the whole metabolic process which is affected by the toxin.

Thus, it is basically false to assume that studies which are carried out neither on man nor animals, represent anything new (or even perhaps methods which have been developed from the pressure of the animal protection debate). For decades, such studies have been part of the normal repertoire of biomedical research. Biochemistry and molecular biology work (almost) exclusively with isolated cells, viruses, bacteria or cell fractions; physiology and pharmacology work mainly with organisms. Alternative tests are therefore neither better nor worse than animal experiments; in only few cases do they directly represent alternatives to animal research. In general, they complement animal research and facilitate the accomplishment of subsequent improved and more detailed work. Moreover these 'alternative' systems of testing help toxicology out of a particular dilemma. Accordingly, most substances to be tested are inevitably required to provide decision scope for toxicological testing; with the quick and cheap so-called alternative tests all the compounds for which the risk of toxicity is low can be found. Expensive test procedures are then performed with these - relatively few - substances. (N.B. All other substances, i.e. those which turn out to be more toxic in screening tests, are discarded and not considered any further). The outlay in time and expensive long-term animal experimentation considers the safety of the long-term therapeutic goals, and serves to answer specialist questions.

In the following section, some of these test procedures will be briefly described. It will become apparent from this discussion that it is more than questionable whether cell culture will ever be as valuable as experiments on the whole animal. Only mutagenicity research is currently performed almost exclusively by cell culture methods, but it is less likely that cell culture will so comprehensively replace whole animal experimentation in other fields of toxicology in the foreseeable future. In view of the dogmatic nature of this statement, one should remember the words of Lord Rutherford on atomic energy: 'The energy produced by the breakdown of the atom is of a very poor kind. Anyone who expects a source of power from the transformation of these atoms is talking moonshine'. It must also be stressed that most so-called alternative methods are currently neither standardised and reproducible nor validated (i.e. in respect of their being confirmed sources of information and relevance for man). For example, bacteria are used in testing for carcinogens ('screening') although they are incapable of getting cancer. To this extent it remains for the future to ensure that a methodological revolution (e.g. in cell cultures methods) does not suffer because of the costs of necessary safety standards (at a time of 'junk foods' and 'junk mail' we should not generate 'junk data', data which is misleading or valueless for risk evaluation). Nevertheless, amidst all the essential criticism of the cell culture alternatives, the fact should not be overlooked that these tests have considerable potential for amassing future knowledge.

Acute toxicity
Thanks to improved cell culture techniques, mammalian cells have been widely used to evaluate the toxic potential of chemicals; primary cell cultures (generally fibroblasts) and organ cultures are preferred because they behave better under different conditions than cell lines, for example (see p. 84). Such investigations at least allow the toxic potential of a series of substances to be evaluated - which would be quite unrealistic in animal experiments. The parameters measured in these studies usually involve growth inhibition, viability, measurement of macromolecule synthesis (DNA, RNA, proteins) and changes of cell morphology; only seldom have cell-specific functions (e.g. liver cell functions) been investigated. This is partly due to the fact that some cell cultures (e.g. of epithelial cells)

are still technically difficult. To date, important studies have been performed on pesticides (e.g. organophosphates), isolated nerves, drugs (especially antineoplastic agents), cosmetics and industrial chemicals, with the result that the few studies in which cell culture and whole animal results (e.g. LD_{50} values) were compared carefully, there was approximately a two-thirds correlation. Because of the complexity of the whole animal (organism-specific, as well as organ-specific metabolic actions, cellular differentiation, hormonal influences, immune system activity and so on) on the one hand, and of the biological simplicity of the isolated cell on the other, it is certainly not surprising that the cell culture alternatives have so far proved inadequate in acute toxicity testing. Cell culture methods can however provide pre-information (in the form of screening tests), and are useful in establishing dosages for the whole animal experiment, thus contributing to the reduction in the number of animals used in research. However, in acute toxicity testing there are no methods suitable for the complete replacement of animal experiments. In toxicology, the aim is to exclude all harmful effects, and a complex test system is therefore required - an isolated organ or cell is of insufficient complexity.

In this connection, it should be noted that studies are currently being performed to determine whether the culture of tumour cells removed from patients can allow the individual sensitivity of patients to cancer chemotherapy to be determined. This is analogous to the behaviour of infectious diseases under chemotherapy.

Special toxicities
1) Inflammatory effects on skin and eyes
Cultures of primary cells and cell lines (generally fibroblasts) are also used in research to replace animal experiments for the determination of the inflammatory effects of chemicals on skin or eyes (e.g. in place of the painful Draize test). The results of these studies may be assessed similarly to those in acute toxicity trials mentioned above. The development of new cell culture methods for the understanding of skin-damaging effects has proved particularly difficult. Although in understanding the phototoxic effects of psoralens for example, a starting point exists (perhaps with cultures of *Candida utilis*), testing agents which irritate and sensitise the skin *in vitro* remains impossible (even though great advances have been made with the development of culture methods for human keratinocytes).

2) Inhalation toxicity
Naturally, few conclusions about inhalation toxicity can be expected from studies of fibroblasts in culture; advances in this field largely depend on the further development of culture methods for epithelial cells. Encouraging reports of the understanding of agents which cause lung fibroses or cancer of the respiratory tract (trachea) are already being discussed. Moreover it should be noted that the lung is a particularly important organ for

immune defence. Studies on cell cultures for the immunotoxicological action of substances are still in their infancy, but have certainly led to important pathophysiological discoveries in inhalational toxicity. For example, the occurrence of tumours after long term inhalation of toxic substances in low concentrations led to the question to what extent overloading of lung clearance and immune defence is responsible for the rat lung cancer caused by diesel exhaust.

3) Neurotoxicity

A comprehensive range of biochemical tests has been developed for the investigation of neurotoxic effects. For example, a very sensitive test for the neurotoxic action of organophosphates relies on the determination of the inhibition of a membrane-bound (synaptosomal) enzyme ('neurotoxic esterase') of the nervous system. Cell cultures however, have only so far been occasionally used to elucidate neurotoxic effects, although the first attempts to culture the spinal cords of frogs were undertaken in 1907. Here too the development of culture conditions which entail the differentiated status of cells and tissues is problematic. In the first series of experiments, either ganglia of motor and sensory neurones, brain in organ culture, and tumour cells from nerve tissue have so far been used. These studies have concluded that, for the determination of neurotoxic effects, cell culture techniques can be considered as auxiliary methods to study the actions of neurotoxic substances, but not as indicators for their *in vivo* toxicity. For this the complex whole animal models (e.g. with intact blood-brain barriers and metabolism etc) are indispensable. Thus, the future here lies more surely in the development of different cell cultures for the detection of cell type and organ-specific effects as preliminary tests for toxicity.

4) Reproduction toxicity

Cell culture systems (of cells, tissues, organs and whole embryos) for the detection of toxic actions on germ cells and of teratogenic effects are quite widely available. The introduction of these *in vitro* tests in routine toxicological testing has however, been relatively unsuccessful. This is mainly due to the fact that there is still no model which covers all the phases of prenatal development (each method only permits the study of a specific section within these phases). Propagation and prenatal development are a meticulously integrated system, and at present, it has been impossible to develop a model which can replace reproduction toxicology studies on the whole animal. However, the importance of cell cultures undoubtedly lies in their role as part of an integrated toxicological test programme, particularly to elucidate mechanisms of action.

5) Immunotoxicity

This specialized area of toxicology is still in its infancy, although there is no doubt that several active substances influence the immune reactions of both humans and animals after acute, but especially too after chronic exposure (often in the lowest concentrations). Soon it must therefore inevitably come to a synthesis of *immuno*toxicology (with knowledge of cell-orientated immunology and of numerous cellular assays for testing immunological functions) and of immuno*toxicology* (with knowledge of the importance of pharmacokinetics and biotransformation etc). The so-called immunotoxicological studies conducted previously often require re-interpretation because of the lack of any such synthesis. Thus, assays of immunological functions are generally insufficient to provide reliable evidence of toxicological effects.

6) Genotoxicity

Today, even if animal experiments for the carcinogenic potential of chemicals are still inevitable, it must still be stressed that in genotoxicity (mutagenicity/carcinogenicity), many cell culture methods have been developed, and their value in detecting chemical carcinogenicity and its molecular basis has been closely scrutinised. Because most chemical carcinogens react with DNA and give rise to mutations, tests for tumourigenic potential based on this principle, on bacteria, yeasts and insects, as well as on cultured mammalian cells, which include interactions of substances with DNA (e.g. binding to DNA, creation of bridging chains, chromosomal aberrations, mutations and induction of DNA repair processes) have been developed. Beyond this, specific procedures are available to study the transformation effect (i.e. transformation of cells in culture into tumour cells after chemical treatment). The reliability of these assays in the determination of carcinogenic potential has been repeatedly checked. Thus it has been established that, by such cell culture assays - presumably it will take not just one, but a battery of these procedures - the *in vivo* carcinogenic potential of diverse chemicals can be accurately established. In addition, substances which do not themselves lead directly to tumours (tumour-promoters and other carcinogenic 'modifiers'), can be determined (at least hypothetically) *in vitro*. By using these assays as screening tests, numerous substances can thus be quickly excluded from further development or application, and it will be possible to reduce the number of research animals (all compounds which cause cancer in humans are also carcinogenic in animal research). However it should be remembered that from cell culture assays too - as from experiments on the whole animal - it has so far only been possible to qualitatively predict the cancer risk for humans. The success of 'alternative methods' in genotoxicity testing is even more remarkable in that this specialty is now the focal point of toxicological considerations. The industrialisation of our society with the consequent flooding of our daily lives with chemicals on the one hand, and the inability to accurately determine the concentration required for carcinogenicity, have forced the increased interest in potential carcinogenic and mutagenic dangers in our society upon science and the general public.

7) Summing up, what does this all add up to?

The 'alternative' new methods in toxicology, and here that means cell cultures, are obviously neither better nor worse than animal experiments, and are only rarely direct alternatives to animal research. Indeed, more often they complement the animal experiment and facilitate the performance of further, more detailed work. As discussed above, toxicology, with its aim of eliminating as much damage as possible, needs a complex test system; thus acute toxic effects cannot be simulated in cell culture, and effects of chemicals on blood pressure, effects on the central nervous system (such as reduced learning ability, balance disorders or giddiness) cannot be studied. Neither, or only seldom, does cell culture identify chronic toxic effects. For example, to date it has been impossible to simulate fibroses or liver cirrhoses (due to alcohol consumption for example) in cell cultures. Thus, cell cultures can never be a complete substitute for the experimental animal.

The human and animal organisms consist of millions of highly specialised cells, which interact with each other directly or indirectly, whereby the balanced interplay of these cells normally means health, but disorder can bring about illness. Tissue cultures involve cells either of the same type, or at best of a small group of cells which interact. If a substance in tissue culture causes no toxic effects, this does not preclude the possibility that it can trigger a devastating effect in the body, since it may develop toxicity due to metabolism or changes during digestion, or because it adversely affects the functional interaction between lungs, heart, blood vessels and kidneys etc. Conversely, a substance which may damage cells in culture, need not be toxic *in vivo* because it is rapidly transformed into harmless metabolites or is quickly excreted.

However, where the animal experiment has shown damage caused by a substance, often only the corresponding cell culture experiment can study it in depth, using biochemical means to explain the molecular bases of the metabolism, the mechanisms of action and the acute toxic and genotoxic effects of chemicals. As a new approach in toxicology, cell cultures are therefore particularly helpful for illuminating partial aspects and are therefore indispensable as complementary methods. Most possible chemical effects and the organism's reaction to them can however only be seen in animal research, and the final evaluation of toxicity (including all data from cell culture studies) is only possible in the animal experiment. At present, the complete renunciation of animal experimentation, for example for the toxicological testing of substances in the human environment, or even merely a reduction in such animal experiments, means an incalculable increase in the risk of possible harmful effects on human health.

4

Methodological Developments

Many animal research methods age so quickly because new physical or chemical methods of detection and new technical developments come into use. Thus, anyone using a method which gives results on the rat which are reliable and relevant to man will not then experiment on the dog. Anyone who can determine a hormone immunologically *in vitro* will no longer measure it by its biological action *in vivo*. Occasionally, like new cars, new methods promise more than they deliver. It is only by carefully comparing the new methods with their predecessors that their value can be assessed. An alternative method is only as good as its safety coefficient. If its reliability can be demonstrated then the alternative method may be accepted and brought into use.

From dog to egg?
Large and small laboratory animals
G. Skofitsch

Our relationship with animals
In discussing the ethical and moral aspects of the use of live experimental animals, a minority, mostly motivated by religious views, or vegetarian animal protectionism, have rigorously expressed the view that all life possesses the same ethical quality and that man has no right to destroy this life. Some philosophers and theologians also plead this 'moral' with an absolutist claim (Gutjahr-Loser 1985), which discriminates on the one hand against a wider form of organic life, namely plant life, and on the other hand, calls for complete conversion to an unnatural way of life. The latter includes a vegetarian diet (the formation of his teeth and alimentary tract show that man is a 'mixed feeder'), complete rejection of all animal products (e.g. leather goods), absolute protection of the animal kingdom, non-interference with fauna and flora (resulting in primeval forest-like grassland-type vegetation) and no treatment of parasitic diseases, etc.

According to the experience of recent years, emotional aversions to animal experimentation are directly related to the type and size and the phylogenetic hierarchy of the experimental animal, if one accepts the premise that together with man, mammals have attained the highest organisational and motivational stages of development. In classifying the animal kingdom, superficial distinctions are made between visibly different animal forms: Single-celled animals, multi-cellular invertebrates, cold-blooded vertebrates and warm-blooded vertebrates. Demands for protection by animal rights campaigners are made mainly for vertebrates (fish, amphibians, reptiles, birds and mammals), and within the vertebrates, for the warm-blooded species (birds and mammals) in particular. For all other animal forms, protection pleas are only raised when they serve to preserve the species (e.g. butterflies). However, even the call for vertebrate protection only seems to be valid when financial and sporting interests do not interfere. How, for example, was it ever possible that fishing for sport and not food, equestrian sporting events which totally exhausted the animals and bullfights etc. were not as vehemently opposed as animal experiments in medical research?

Sensitivity for animals depends on ties of feeling and possibly of social bonds (intimate relationships); through these it is possible to lift a particular animal (individual) out of collective anonymity and to converse with it (Gartner 1985). 'Intimate relationships' are easy to develop with dogs, pigs, cats and monkeys, but are much more difficult with hamsters, rats and mice. Accordingly, from statistical enquiries, sentiments about the death of an animal can be placed in the following order:

monkey > dog > cat > goat > sheep > rabbit > pig > guinea-pig > hamster > rat > mouse

or simply be represented as:

'talent for communication' > economically useful animals > pests (Gartner, 1985).

Even amongst 'communication talents' a distinction must be made between socially recognised 'sporting animals' and 'useful animals'. Pigs for example, in their intelligence, ability to learn and to be trained, play instincts and attachment to people, are like dogs. Thus, pig dressage in the circus and the use of pigs trained to find truffles are not sensational. However, who would want to take so intelligent and well trained a pig for a walk on a lead in a public place? The pig's place in society is amongst the 'useful animals' and not the 'pets' - irrespective of its intellectual abilities.

Anthropomorphic considerations have also found their way into the different recommendations and guidelines for animal research. So for example in the handbook on the care and maintenance of experimental animals, published by a university animal protection society (Worden, 1947), there are three categories of privileged animals:

1. Horses, donkeys and mules may never be used for animal experiments if the experiment can be undertaken on any other species of animal.

2. Cats and dogs may not be used for an animal experiment without anaesthetic if the experiment can be undertaken on any other species of animal.

3. Monkeys, rodents and all other vertebrates enjoy general protection and may only be used for research if the research cannot be carried out on an invertebrate or lower organism.

4. Invertebrates have no protection.

This distinction of 'privileged' and 'non-privileged' animals has persisted into today's animal protection and animal protection legislation. On these grounds, the use of ever smaller, lower organisms for animal research is seen as the legitimate alternative to the large animal for research. This development is desirable, not least from the economic standpoint.

Pure research - applied research
Basically, a distinction must be made between two different lines of research, namely pure and applied research. In general, pure research has no directly discernible connection with man and his health. Its perceptions are purely scientific in character and spring from the need for man to study his environment. The results cannot immediately or directly be evaluated in the medical or technical context. Pure research is therefore designated as non-targeted research. On the other hand, applied research is directly for man's benefit, for example in medical or pharmaceutical experimentation.

In spite of the lack of directly applicable data from pure research for man's benefit, neither the justification nor the importance of the purpose of this research for the pursuit of scientific knowledge should be denied. Pure animal experimental research accounts for about 5% of the higher vertebrates used in animal experiment (Creutzfeldt, 1985). Most experiments are performed on lower organisational forms, isolated organs or cell cultures in order to reduce the number of complicating factors, or the knowledge is obtained from the observation of animal behaviour. The most important points of interest are (according to Ullrich, 1985): studies of the building blocks of living matter, their functional interdependencies, the ensuring of continuity (cell division and propagation) and life expression (ethology, behavioral research). Pure research is concerned with all forms of organic life, and does not therefore restrict itself to any animal species and fully accepts alternative methods. Applied research is based more on the general knowledge of biological interactions which have been promoted by pure research.

'Applied' or 'purpose-orientated' research, however, is quite different. Here there is a specific goal, which can be achieved in several ways and with different experiments. In medical, biological and biochemical research there are two primary goals: (1) to simulate human diseases in animal models to test the action of new, previously untried drugs (see p. 15), and (2) to establish the tolerance (see p. 39) of substances in an animal organism (toxicity testing). To these ends, animal models are developed by surgical, pharmaceutical or genetic methods, and are used to infer the action of a substance in the human organism. Furthermore, animal experiments assist the development of surgical methods (see p. 197).

More highly organised animals in applied research
In the early days of medical research, technical reasons dictated the use of easily maintained, but large and sturdy experimental animals. At the beginning of this century, animal experiments for medical research mainly involved dogs, cats and rabbits. It was argued that these animal species are developmentally closer to man than the lower vertebrates or invertebrates, and it was expected that their reactions would be similar to those of humans. Surgical techniques were restricted (no microsurgery), and biological measurement data could be elicited more easily with the relatively simple apparatus of the time. There were no physicochemical detection methods for the little-known hormones, so they had to be isolated from large amounts of tissue or body fluids (e.g. blood and urine) and tested in animal experiments for their biological action (Jackson, 1917; Sherrington, 1919; Haberland, 1926; Burn, 1948).

In the last few decades, the development of more sensitive analytical procedures and electronic measuring equipment have enabled the transfer of functional investigations, which were originally performed on dogs or cats, to small rodents, especially the rat (Farris and Griffith 1942; D'Amour and Blood 1948). This has the advantage that these animals can be widely bred in genetically homogenous strains so that their maintenance and feeding is optimized. Every subsequent publication should state the strain of experimental animal used, so that the same results can be obtained in comparable studies at different institutions. Thus, a great saving in the numbers of experimental animals is possible.

About one third of animals used in applied research are used for medical-pharmaceutical research, partly in standard physiological or pharmacological tests - 80% of which are carried out on rats and mice (Kummerle *et al.* 1984) - and partly in anatomical and surgical experiments or for the simulation of disease. The standard physiological-pharmacological tests cannot be transferred to lower animals and organisational forms. Anatomical and surgical procedures must also be performed on animals whose surgical anatomy corresponds closely to that of man (pigs and rabbits are the main surgical research animals). However, for the simulation of diseases using animal models to study new therapeutic potentials, increasingly small (and lower) animal species are used. Thus, in 1952 a model was developed to study hypertension treatment and cardiac muscle insufficiency using surgical methods in the dog (Barger et al. 1952). This was followed by similar models in the cat (Spann et al. 1972), the guinea-pig (Schmidt et al. 1977) and the rat (Hutt et al. 1980). Today, specific strains of rat are available for this kind of research, in which hypertension is genetically determined. Thus, drugs for the control of blood pressure can be tested without any of the earlier surgical manipulations on experimental animals.

Lower animals in applied research
Invertebrates and lower vertebrates were used in early pharmacological research, but knowledge of them was regarded rather as a curiosity: single-celled life forms (protozoa) displayed reactions to different drugs. The action of strychnine was tested on crabs, and that of curare on beetles. However, the modes of action of these substances could not be extrapolated to humans.

Pure research on lower animals often yielded unexpected significant knowledge about vital human processes. Thus for example, the Nobel Prize winner Otto Loewi used frog hearts to show that a nerve triggers a stimulus to the heart muscle, not via a direct electric impulse, but by releasing a chemical from the nerve endings (see p. 8). This principle is generally applicable for the stimulus transfer of nerves to reacting organs. The Danish pharmacologist Skou isolated an enzyme (potassium sodium adenosine triphosphatase) from the nerves of squid which was later found in the heart muscles of vertebrates. It is now known that this enzyme is vital for the effect of cardiac glycosides (digitalis). The Italian pharmacologist Erspamer studied components of toad skin and discovered a series of compounds, which were subsequently found in the human intestine and which fulfil important roles as hormones and neurotransmitters. Some of these compounds are involved in the transfer of pain stimuli to the brain and in the provocation of asthmatic attacks.

These examples should illustrate the close involvement of pure 'academic' research and the practical, medical uses of this knowledge.

Safety investigations
About two thirds of the animals used in applied research are used for legally prescribed safety investigations. Such studies are required not only for drugs, but also for skin care products, detergents and cosmetics, as well as for many other substances which are found in the household and at work. In this way the non-toxic properties of a product at the normal commercially used dosage or concentration are discovered. However, this

presupposes the awareness of doses which are actually harmful. Monkeys, pigs, dogs, rabbits, guinea-pigs, rats and mice are used for these tests (Ther, 1965). Because of the relatively high number of animals used to ensure that the results are statistically valid, these experiments are particularly disliked by the public. At a recent international meeting (International Conference on Practical *In Vitro* Toxicology 1985), alternative methods to the animal experiments used previously were discussed and evaluated. However, the development of alternative methods presents us with new problems. Which methods are really 'alternative'? Are these methods as reliable as the conventional techniques? Will alternative methods win legal recognition or not?

Arising from the suggestion that higher experimental animals should be replaced by lower life forms, it can be established that, both in research and in safety tests (certainly on economic grounds), the smallest possible (lowest) animal which satisfies the aims of the investigation should be used (Scheuch, 1986). Frequently, the progression 'from dog to rat' is perceived to be no more than an adequate 'alternative', and the withdrawal of all animal experiments and the development of isolated organs, cell cultures and computer systems is demanded.

The use of computers as a substitute for animal research in safety studies is limited (see p. 106); isolated organs and cell cultures do not allow for the reciprocal action of different organ systems of a higher life form, unless they are used for very defined problems.

Experiments on chicken embryos are currently being promoted because they represent a borderline area between *in vitro* and *in vivo* investigations. At the same time they do not contravene any ethical and legal aspects such as the animal experiment laws. With this method, test substances are no longer injected into an animal, but are introduced into incubated eggs through a 'window', and the development of the embryo is then observed. The value of this test is limited, in the opinion of most research groups, because although the system can differentiate between extremely toxic and non-toxic substances, it is unable to account for the individual resorption barriers and detoxification mechanisms of the mammalian organism and these still have to be elucidated in animal experiments.

For the investigation of embryotoxic and teratogenic effects (damage to the developing life and genes) test substances are administered to pregnant dogs, rabbits, rats or guinea-pigs, and the effects on fetal development are examined. An 'alternative method' arises from this, namely that mammalian embryos can be grown in an incubator under standardised conditions after surgical removal from the mother animal. The test substance then no longer needs to be given to the mother, but is put directly into the culture medium containing the live embryo. However, how far this is an 'alternative method' is open to question.

Another alternative in teratogenicity research is the hydra test. Here the regenerative capacity of lower organisms is put to good use. If a hydra is cut into pieces, each piece is normally capable of growing into a viable creature (regeneration). If a teratogen is placed in the incubation medium containing the hydra piece, the regeneration capability will be impaired, giving an indication of the toxicity of the test compound.

The layman will be fascinated by accounts of such newly developed test procedures and will ask why these developments were not used immediately. The responsibly aware expert however, knows that conventional methods have been developed, improved and tested over several decades and the behaviour of the test system is therefore well established. He also knows that establishing safety limits begins with the evaluation of the reliability of the process used for such an investigation. An alternative method must therefore be compared with the existing methods over a long period, in order to be certain that the new method is completely comparable.

Nevertheless, it has finally been established that some animal experiments can be supplemented and reduced by new and alternative methods. Animal experiments themselves, as the final safety assessment before clinical trials on humans, are however, presently irreplaceable. Our qualified hope rests in the continuing development and testing of new alternative processes which can, in the future, render further animal experiments superfluous.

References

D'Amour, F.E. and Blood, F.R. Manual for laboratory work in mammalian physiology. Univ. Chicago Press 1948.

Barger, A.C., Rohe, B.B. and Richardson, G.S. Relation of valvular lesions and of exercise to auricular pressure, work tolerance, and to development of chronic congestive failure in dogs. Amer. J. Physiol. 169 (1952) 384-399.

Burn, J.H. The Background of Therapeutics. Oxford Medical Publications. London 1948.

Creutzfeldt, O.D. Ethik, wissenschaft und tierviersuche. In: Ullrich, K.J. and Creutzfeldt, O.D.: Gesundheit und Tierschulz. Econ. Düsseldorf 1985 (S. 11-43).

Farris, E.J. and Griffith, J.Q. The Rat in Laboratory Investigation. Lippincott, London 1942.

Gärtner, K. Ist der derzeitige diskussion über tierversuche durch unsere aktuellen lebenqualitäten bedingt? Zur mensch-tier-beziehung aus historischer und sozialempirischer sicht. In: Ullrich, K.J. and Creutzfeldt, O.D.: Gesundheit und Tierschulz. Econ. Düsseldorf 1985 (S. 94-105).

Gutjahr-Loser, P. Die argumente der tierversuchsgegner. In: Ullrich, K.J. and Creutzfeldt, O.D.: Gesundheit und Tierschulz. Econ. Düsseldorf 1985 (S. 106-122).

Haberland, H.F.O. Die operative technik des tierexperimentes. Springer, Berlin 1926.

Hutt, P.Y., Rakousan, K., Gastineau, P., Laplace, M. and Clousard, F. Basic Res. Cardiol. 75 (1980) 105-108.

Jackson, D.E. Experimental Pharmacology. Henry Kimpton, London 1917.

Kümmerle, H.P., Hitzenberger, G. and Spitzi, K.H. Klinische Pharmakologie. Ecomed. Landsberg 1984.

Sherrington, C.S. Mammalian physiology. A course of practical exercises. Clarendon Press, London 1919.

Scheuch, E.K. Das tier als partner des menschen in der industriegesellschaft. In: Studium Generale: Mensch und Tier. Bd. III/IV, Schaper, Hannover 1986.

Spann, J.F., Cowell, J.W., Eckberg, W.L., Sonnenblick, E.H., Ross, J. and Braunwald, E. Contractile performance of the hypertrophied and chronically failing cat ventricle. Amer. J. Physiol. 223 (1972) 1150-1157.

Ther, L. Grundlagen der experimentellen arzneimittelforschung. Wiss. Verlag. Stuttgart 1965.

Ullrich, K.J. and Frömter, E. Tierversuche in der biologischen grundlagenforschung. In: Ullrich, K.J. and Creutzfeldt, O.D.: Gesundheit und Tierschulz. Econ. Düsseldorf 1985 (S. 125-139).

Worden, A.N. The UFAW Handbook on the care and management of laboratory animals. Bailliere, Tindall and Cox, London 1947.

Tiere in der arzneimittelforschung. Bundesverband der pharmazeutischen industrie, 1986.

Tierversuche in der diskussion. Bundesverband der pharmazeutischen industrie, 1986.

Experiments on the organs of dead animals
The isolated organ
P. Holzer

The various levels of biomedical research
Biomedical research avails itself of various methods of investigation. Together with animal experiments (research on the living animal), three other groups of research methods are currently available which are not performed on living animals. These techniques are often imprecisely called 'alternative methods'. The first group includes experiments which take place not on the whole animal, but on isolated organs, tissues, cells or cell components. For these to be obtained it is usually necessary for an animal to be killed, either in the laboratory or in the abattoir. Tissues which are surgically removed from man or animals can be considered as an exception to this rule (see p. 75). In some cases however, cells (e.g. blood and liver cells) may be procured without killing, or even without permanently damaging the organism. The second group of methods concerns the use of tissue cultures, whereas the third group includes all experiments using 'non-living matter', such as purely

physical or chemical investigations, and the use of computers to simulate biomedical procedures. The last two procedures are discussed in other chapters of this book.

Whether these methods actually represent alternatives to animal experiments can only be determined in conjunction with the aims of biomedical research. The chief object of fundamental biological research is to clarify the ways in which cells are formed and work together to facilitate the continuance of life and life processes (see Ullrich and Creutzfeldt 1985). In close cooperation with basic research, practical medical research attempts to probe the causes of disease (disorders of specific life processes) and to devise ways of curing them. The organism of man and of most animals is made up of cells which form tissue complexes (e.g. muscle and fat tissue) and organs (e.g. heart and brain). Complicated regulatory mechanisms determine the 'purposeful' interaction of cells and organs: it is primarily through this that an organism can survive. Biology and medicine try to study these processes on three levels. Thus, the investigation of biological functions, and the mode of action of drugs and other therapeutic measures are respectively performed:

a) on the level of cells and cell components
b) on the level of organs which consist of many different cells working together in specific ways (e.g. heart, liver, stomach and brain)
c) on the level of the organism, which consists of different organs whose all-important cooperation ensures that man can, for example, hear, see, move and think.

For each level of research, appropriate investigative techniques are required. Blood pressure and its regulation, or the action of a drug on blood pressure can only be examined on the whole organism, and not on an isolated artery. In other words, every scientific problem can only be solved by specific methods, each of the three defined research levels requiring a targeted scientific procedure. Thus, it is clear that animal experimentation is the only adequate method for studying the life processes of an organism. Investigations with isolated organs or cell cultures may be considered only as supplementary methods, rather than as alternatives. On the other hand, experiments with isolated organs represent a suitable approach to studies at the organ level, and it is the purpose of this chapter to discuss the place of these methods in the biomedical sciences.

A look at the past
Although investigations with isolated organs had been performed in the 18th century, decisive progress in these techniques was first achieved towards the end of the 19th century (Magnus 1904, Holmstedt and Liljestrand 1963, Paton 1984). These advances depended primarily on the discovery of the optimum conditions under which an isolated organ can survive for a sufficient length of time. Survival is only possible in a nutrient solution which contains specific essential ions and nutrients and has a specific pH value.

The optimum composition for this solution varies slightly for different organs of the same species, and for the same organs of different species. The further development of the methodology for isolated organs, tissues, cells and cell cultures was thus very closely associated with the development of appropriate nutrient media (Paton 1984).

Another problem arose in the provision of oxygen to the isolated organs, since an isolated organ only remains capable of activity with an adequate supply of oxygen. While it is sufficient for small and thin-walled organs (e.g. frog heart or guinea-pig stomach) to perfuse the nutrient medium with oxygen, such a procedure is quite inadequate for larger and/or thick-walled organs. In these cases it is necessary to perfuse the vascular system of the organ with oxygen-containing nutrient solution. The development of the perfused isolated organ technique mainly facilitates the study of isolated large organs such as mammalian heart, liver, lungs and kidneys (Paton, 1984).

Potentials and limitations of the methodology of isolated organs
The methodology of isolated organ culture has now become an established component of biomedical research which cannot now be repudiated. Further technical developments have made almost all organs and tissues suitable for these procedures. Two examples should be briefly mentioned. In suitable conditions, the isolated heart (mainly of rats or guinea-pigs) beats for many hours, and is therefore an extremely useful model for testing cardiac agents (see p. 118). In the isolated intestine, many, but not all components of the digestive process can be detected, including intestinal motility, the products of digestive enzymes and the assimilation of digested foodstuffs.

The use of isolated organs has contributed to the decline of experiments on living animals both in basic research and in the investigation, testing and production of drugs. Amongst the *scientific* advantages, *economic* and *ethical* considerations were also crucial as well as the framework of *legalistic* conditions for the increasing use of isolated organs.

The *scientific* advantages of isolated organ techniques were soon recognised. R. Magnus describes them in his first work on experiments on the isolated intestine in 1904: "During research into the activities (phenomenon of motility) of individual organs it has always proved advantageous, in addition to experiments on whole animals with their complicated innervation processes and the changing influences of circulation, to carry out investigations on surviving organs in order to establish which activities this organ is capable of on its own and unaffected by any external factors". Whilst the status of the methods used in basic research is considered here, their particular significance for the *development of drugs* should also be noted from the following two examples:

1. In the absence of appropriate chemical detection procedures, many drugs can only be quantified by their biological actions (e.g. contractions of intestinal or vascular musculature). In these 'bioassays' a suitable isolated organ is generally much more sensitive and accurate than the whole animal.

2. In the development of many drugs which act on the central nervous system (e.g. pain-killing drugs) certain isolated organs have been shown to be extremely useful models for the complex central nervous system (see pp. 174 and 179). Strong pain killers (analgesics) such as morphine inhibit contractions of isolated intestinal segments or of the isolated spermatic duct via the same cellular mechanism of action, through which they relieve pain in the central nervous system. Morphine however, carries two serious side effects, namely respiratory inhibition and the onset of addiction, which indicates the need to develop more selective analgesics. Experiments on isolated organs have now succeeded in differentiating cellular mechanisms of action for pain relief from those causing side effects. This advance justifies the hope of developing selective analgesics without these side effects (Tyers 1986). This example should make it clear that a major proportion of drug development on isolated organs, tissues and cells will ensure that only the most promising compounds finally have to be tested on the living animal (Tyers 1986).

Isolated organ techniques cannot however completely replace an experiment, at least not in every case, where it involves the elucidation of reactions between the various organs in the whole organism. In addition, during drug development, research on the whole animal is only important when the assimilation, distribution and excretion of active compounds, or their effects on blood pressure and circulation or on the central nervous system have to be established. Here the limitations of isolated organ techniques become clear. A further restriction arises from the question of whether isolated organs, tissues or cells behave very differently from the way they do in the whole organism. Such considerations strengthen Magnus's view quoted above, that studies on isolated organs should always be supported by experiments on the whole animal. This requirement can be upheld by impressive examples from the history of medicine (see pp. 8, 11 and 21).

Closely associated with the scientific viewpoint, *economic* considerations are also responsible for the increasing use of isolated organs, tissues and cells. These methods are generally considerably cheaper than experiments on whole animals. For example, between 1977 and 1984, the transition to isolated cells, cell components and cell cultures produced a 40% drop in the use of experimental animals in the Federal Republic of Germany alone (Federal Union of the Pharmaceutical Industry 1986). Although the procurement of isolated organs is not possible without killing an animal, their use guarantees an economic exploitation of animal life. Thus for example, 20 portions can be obtained from the small intestine of a guinea-pig, each of which can be used as a model for the whole intestine: in place of one experiment on the whole animal, 20 different tests can be carried out (Paton 1984). In addition, not only the intestine, but also the other organs of the dead animal are simultaneously available for experiments.

In the evaluation of 'alternatives' to animal experiments, two aspects are currently of importance. *On the one hand* it is beyond question that every unnecessary animal experiment should be avoided and that appropriate international and national legislation is passed and adapted to the current status of scientific knowledge. In many cases, particularly in drug development, the isolated organ techniques assume an outstanding importance in the reduction of animal experiments. *On the other hand* it must be clearly stated that not all animal experiments can be replaced by other methods. Considered scientifically, animal experiments and studies on isolated organs, tissues and cells have different, mutually complementary aims and can therefore not be arbitrarily interchangeable. For medicine, a total view of the human body (comprehensive medicine) is often demanded, and rightly so: it is difficult to see how such a goal can be attained without research on one complete model, such as is represented by the whole organism.

References

Bundesverband der Pharmazeutischen Industrie e.V. Pro & contra tierversuche, argumente, dokumente. Frankfurt am Main 1986.

Holmstedt, B. and Liljestrand, G. Readings in Pharmacology. Pergamon Press, Oxford 1963.

Magnus, R. Versuche am überlebenden. Dünndarm von Säugethieren. 1. Mittheilung. Arch. Physiol. (E. Pflügur) 102 (1904) 123-151.

Paton, W. Man & Mouse. Animals in medical research. Oxford University Press. Oxford 1984.

Tyers, M.B. The use of isolated tissue preparations in the discovery of drugs for mental disease and pain. ATLA 13 (1986) 123-151.

Ullrich, K.J. and Creutzfeldt, O.D. Gesundheit und tierschultz. Wissenschaftler melden sich zu wort. Econ. Düsseldorf 1985.

Offal for research?
Organs from the slaughter house
E. Beubler

As a boy, Salomon van Zwanenberg may have harboured the wish to become a doctor, but he finally remained what his father was; a pig butcher. It may however, be connected with his childhood dream that he saw the numerous pigs which were slaughtered daily, not only profitable as food: he also saw potential in the organs which were treated as worthless offal. He may also have known that elsewhere, all products of dried glands which apparently had health-giving effects, had a commercial use.

Together with the Amsterdam pharmacologist Ernst Laqueur, he founded a company in 1922. The pig butcher undertook the financing and Professor Laqueur the scientific management. The first enterprise was the large scale production of insulin from the pancreas. Today, the company they founded belongs to the biggest drug manufacturers in the world.

Since then, the value of slaughter house waste has not always been so spectacular. However, a whole series of biological substances was first discovered in biological material from abattoirs, which was, and still is, used therapeutically.

As examples, only the discovery of substance P in horse intestines (1931) and thyrotropin-releasing factor in sheeps' brains will be mentioned here. 250,000 sheep hypothalami had to be used to obtain 1 mg of the latter compound, while 500 kg of suprarenal gland from 20,000 animals were used to obtain 1 g of the steroid hormone corticosterone (see p. 109). For the discovery of serotonin (1947), 900 litres of serum, obtained from 2 tons of cattle blood was needed. At the time, researchers were unaware that their work was favoured by the fact that cattle serum, compared to that of laboratory animals, has a particularly high serotonin content.

The expression 'offal disposal' is not quite appropriate if one considers that, for one gram of vasoactive intestinal polypeptide, a highly active peptide from pig intestine worth around 1.5 million DM, approximately 5,000 kg of waste intestine has to be processed. Taking the calculation further, it appears that pig 'waste' is scientifically very valuable, although this does not consider the costs of obtaining these biological compounds. It should merely be noted that, in research, slaughter house material should always be used if possible.

For the investigation and subsequent production of hormones, peptides, and enzymes, amongst other biologically-active compounds, cheap material from abattoirs is certainly an inexhaustible source, since these compounds are mostly found in organs which are unsuitable or worthless as food. Apart from the ethically acceptable use of this material, a product obtained from herds of animals is more economical than one from a special laboratory animal, just as a food mixer costs only one tenth of a laboratory homogeniser.

Slaughter house products also find their uses in therapy. Insulin is still obtained in large quantities from pig pancreas. So-called 'human insulin' is likewise now produced from pig insulin by chemical transformation (amino acid exchange). Although the use of cattle spleen extract for the treatment of 'splenic obstruction', as described by Paracelsus, as well as the use of dried cattle thyroid (*Glandulae Thyreoideae siccatae*) for thyroid deficiency is obselete, other organs or secretions like ox gall and pancreas extract are still used in numerous multi-ingredient preparations for digestive ailments.

Surgery also uses biological material from slaughter houses. Surgical sutures ('cat gut') is produced from the sheep intestine, and heart valves of pigs and sheep serve as so-called biovalve prostheses in humans. Parts of pig skin are used as interlacing collagen fibres to stop bleeding from parenchymal surfaces of liver or kidneys. Pig skin can also be used for burns. Even deproteinised pig bones are used as a matrix for human bone defects.

The use of organs from abattoirs as pharmacological and physiological models is more difficult and therefore limited for various reasons. Slaughter houses are often so far away from research laboratories that it is impossible to bring a fresh organ at the right time to the appropriate apparatus. For example, a guinea-pig heart must be perfused with nutrient solution and oxygen within 90 seconds of the animal's death. It then remains capable of functioning for several hours. The chemicals used for such an experiment are of the highest purity and are therefore extremely expensive. Organs from slaughter houses are considerably larger than those of laboratory animals, and thus an experiment on an isolated pig or ox heart would, quite apart from the technical difficulties, cost enormous sums of money. The perfusing of an isolated adrenal gland from a cow (30 g) is easy to accomplish, while the same procedure in the rat adrenal (30 mg) is only achieved with difficulty.

The cow's eye too, because of its size, is easier to use as a model than the eye of a laboratory animal. Thus, zonulolysis, a method for enzymatic resolution of the zonular fibres which facilitates the careful removal of the lens in cataract operations, has been successfully developed on cows' eyes.

The coronary vessels of a cow can readily be studied in the laboratory since smooth muscle organs tolerate the journey from slaughter house to research institute well.

Tracheal musculature is also the subject of numerous investigations in asthma therapy. However, it was discovered that the tracheal musculature of the cow is exceptionally insensitive to methyl xanthines (e.g. to theophylline, an important agent in the treatment of asthma). On the other hand, the guinea-pig is the ideal research animal since its tracheal musculature is similar to that of man.

The spleen and lung arteries, as well as various veins and other smooth muscle organs are used wherever possible for resolving biochemical, physiological and pharmacological questions. With these the researcher allows for the fact that certain animals are only slaughtered on certain days, and that the organs must be transported in special containers with nutrient solution as quickly as possible.

For certain problems involving coronary vessel diseases, endothelial cells (the inner layer of cells in blood vessels) of cattle aorta are currently being used with great success. These cells can be grown *in vitro* where they constitute an ideal research model in cell cultures.

Human surgical and postmortem material
Ethically justifiable?
H. Denk

The aim of biomedical research is to elucidate the pathogenesis of human disease for its prevention, diagnosis and therapy. Chronic diseases in particular, for example arteriosclerosis and malignant tumours, are of primary interest. A central problem is the extrapolation of experimental data, often from animal experiments, to sick humans. This causes difficulties, particularly due to specific differences between man and experimental animals, and individual, genetically-based, variations, but also because of environmental factors. An additional possibility (although not always adequate), lies in establishing the reactions of the whole organism to toxins or drugs using tissue obtained from operations or postmortems. This is frequently performed by morphological tests (light or electron microscopy, immunohistochemistry, *in situ* hybridisation etc.), which permit the correlation of morphologically detectable changes with anamnestic data (e.g. uptake of drugs, exposure to industrial toxins or environmental influences etc.) and the clinical findings. This in turn allows the developmental progress of organ damage (pathogenesis) as well as potential therapies to be studied. Although such investigations can stimulate experimental studies on cultured cells or tissues, or on animals, they cannot replace them. Human surgical and postmortem material is presently used, however, for obtaining various medically useful substances (e.g. antigens for immunisation and hormones). Cultured human tissue and cells form a bridge between animal experimentation and clinical medicine. The development of useful human cell lines however, is still in its infancy, and thus animal systems offer important methodological foundations.

Morphological investigations on operatively, bioptically and autoptically obtained human tissue and organs
Morphological investigations on operatively and bioptically obtained human tissue are mainly used for diagnosis. They aim to recognise, classify or even exclude pathological changes as a basis for the understanding of pathogenetic and epidemiological connections, therapy and prognosis.

In addition to classic light microscopy of stained organ sections, immunohistochemistry (the demonstration of tissue-bound antigens using specific labelled antibodies) and electron microscopy (to demonstrate ultrastructural changes) are also used. However, the picture of disease revealed by this approach is only a snapshot in time, which moreover can depend on a number of variables (individual fluctuations, environmental conditions etc.) and therefore can only be generalised by studying a large number of cases.

Human tissue as the starting point for cell cultures
Cell and tissue culture offer worthwhile, versatile and proven biomedical research systems, mainly for the study of metabolic processes, diagnosis of genetic defects, understanding pathogenetic associations, hormone action and the determination of toxic, mutagenic and carcinogenic effects. Ideally, the culture conditions should guarantee a milieu which facilitates the demonstration of cellular and tissue function in a controllable *in vitro* system. A prerequisite is to obtain culturable (surviving) human material from operations and from postmortems as soon as possible after death. The use of surgically removed human tissue for experimental investigations always requires aware and responsible collaboration of the disciplines involved, since, by using tissues for culture purposes neither the treatment of patients nor the pathologist's diagnosis must be prejudiced. The transferability of the results gained by this system to the *in vivo* situation depends on the characteristics of the cultured cells (Harris *et al*. 1980). It should be noted that by isolating cells from tissue and subsequently growing them in the test tube (*in vitro*), important cellular properties, for example reactivity to hormonal stimulation via hormone receptors, and membrane and cytoskeletal structure, can be altered. Methods currently exist for culturing cells from lung, bronchus, vascular endothelium, myocardium, skin (epidermal cells, melanocytes, fibroblasts etc.), uterus, breast, placenta, endocrine organs (e.g. pancreatic islet cells), gastrointestinal tract (including the liver) and the urogenital system (kidney, bladder and prostate). Many organ explants can now be maintained in culture for several months, but some (e.g. intestinal mucosa) for only a few hours (Harris *et al* 1980). Permanent human cell lines are derived mainly from tumour tissue, but they have only a limited resemblance to normal human cells and tissues.

Cultivated amniotic fluid cells (taken during amniocentesis) can be used for the prenatal diagnosis of genetic defects (Bresch and Hausmann 1972). These cells can be tested for their chromosomal make-up (e.g. for chromosomal aberrations) or tested for biochemical defects. *In vitro* experiments also play a role in the maintenance of organs to be used for organ transplantation (e.g. kidneys).

Human tissue as the starting material for the production of antibodies for diagnostic and therapeutic purposes

A prerequisite for antibody production is the obtaining of corresponding antigens from normal or pathological (e.g. neoplastic) tissue, which are then used for the manufacture of polyclonal or monoclonal antibodies. Antibodies against cell-specific substances, e.g. cytoskeletal proteins, are used for the histochemical classification of various poorly differentiated tumours with an uncharacteristic morphological appearance, but which have different implications for therapy. The characterisation of hormone-producing tumours is immunochemically possible using antibodies directed against the appropriate hormones (Mach *et al* 1980, Polak and van Noorden 1983, Yelton and Scharff 1981). Antibodies against tumour-associated (e.g. oncofetal) antigens can be used for their detection in tissue and serum, and are therefore diagnostically important. Such antibodies are also used, after combination with radioisotopes, for tumour localisation, and in combination with cytostatic drugs, for specific tumour therapy (Polak and van Noorden 1983, Yelton and Scharff 1981). Of particular practical importance are antibodies directed against leucocyte surface antigens, which may differentiate cell types in lymphomas and other haematological malignancies. The yield of highly specific antibodies can be considerably increased by hybridoma technology for the production of monoclonal antibodies, without increasing the number of animals to be immunised (Bhan 1984, Diamond and Scharff 1982, Köhler and Milstein 1975). For the production of monoclonal antibodies, the spleen cells of immunised animals are fused with myeloma cells which then produce large quantities of antibodies *in vitro* (Köhler and Milstein 1975; see also p. 159).

Isolation of active substances (e.g. hormones) from human tissue

Therapeutically important substances, e.g. growth hormones from hypophyses, can be obtained from human postmortem material.

Operations on cadavers

Operations on cadavers are an important requirement for the development of new surgical procedures in the interest of the living and are therefore ethically acceptable. They are performed like postmortems and entail the removal of organs from the body (see below). Operations on research animals can be reduced in this way, but not replaced, since the success of the operation largely depends on the functioning result and thus requires the functionally intact organism. This is particularly important in transplantation in which the surgical technique, together with preparation and functional maintenance of the transplanted organ, is only one contributory factor to success - albeit the most important.

Removal of human tissue: legally and ethically acceptable?

According to existing Austrian law, the removal of organs and parts of organs from cadavers is permissable if the patient (or the legal representative of a patient who is not answerable for himself) has not prohibited the removal of his organs before his death (Holczabek and Kopetzki 1986). Express permission of the deceased (or his representative) is not needed. The opposition of relatives who are not the legal representative is discounted. Organs and parts of organs of the deceased may not, however, be the subject of transactions for profit. Organ removal may not be the cause of any offensive deformation of the corpse. Organ removal must be documented and is accompanied by medical discretion. With these prerequisites, the removal of organs and parts of organs is ethically acceptable for purposes of healing (transplantation), diagnosis, teaching and research, and is subordinate to the greater good - 'life'. Apart from obtaining organs for transplantation, organ removal is carried out within the framework of postmortems, which, in line with existing regulations, are performed in accordance with supervised sanitary or legal orders, or the preservation of public or scientific interests (e.g. doubtful diagnosis or investigation of surgical failure). On the other hand, the removal of a healthy organ from living patients without their legal consent constitutes the criminal offence of causing actual bodily harm.

Humans as 'guinea pigs'?
The aim of biomedical research is to understand human diseases as the basis for prevention and healing. However, on ethical grounds, man himself can serve as the object of research only to a limited extent. On the other hand, pathology can promote worthwhile knowledge through the study of surgical and postmortem material. Thus, in terms of epidemiological research, a correlation can be established between discernible noxae (environmental conditions, conditions at work, smoking, drugs etc.), and the appearance of organ damage. However, this approach cannot determine an exact cause-effect relationship, and definite conclusions about a connection are only possible if the changes have occurred in a large number of people. Testing drugs by clinical trials on patients is certainly necessary to confirm the effect, but is only acceptable after all other tests have been used to minimise the risk. It is most successful if it is used first on volunteer healthy subjects and finally on sick people. Self-testing, because of the small number of test subjects available, together with individual variations, rarely leads to good results.

References
Bhan, A.K. Application of monoclonal antibodies to tissue diagnosis. In: De Lellis, R.A. Advances in Immunohistochemistry, New York 1984.
Bresch, C. and Hausmann, R. Klassische und molekulare genetik. Springer, Berlin 1972.
Diamond, B. and Scharff, M.D. Monoclonal antibodies. J. Amer. Med. Assoc. 248 (1982) 3165-3169.
Harris, C.C., Trump, B. and Stoner, G.D. Normal human tissue and cell culture. In: Methods in Cell Biology, Vol. 21, A, B. Academic Press, New York 1980.

Holczabek, W. and Kopetzki, Ch. Rechtsgrundlagen von organtransplantationen. Weiner klin. Wschr. 98 (1986) 417-420.

Köhler, G. and Milstein, C. Continuous cultures of fused cells secreting antibody of predefined specificity. Nature 256 (1975) 495-497.

Mach, J.P., Buchegger, F., Formi, M., Ritschard, J. et al. Use of radiolabelled monoclonal anti-CEA antibodies for the detection of human carcinomas by external photoscanning and tomoscintigraphy. Immunol. Today 2 (1981) 239-249.

Polak, J.M. and Van Noorden, S. Immunocytochemistry. Practical applications in pathology and biology. Wright PSG, Bristol 1983.

Yelton, D.E. and Scharff, M.D. Antibodies: a powerful new tool in biology and medicine. Ann. Rev. Biochem. 50 (1981) 657-680.

What can we learn from cells?
Cell cultures and their limits
H. A. Tritthart

The smallest living systems and the building blocks of our body are known as cells. They are often very different in form and function, although their elements (nuclei, intracellular structures, fibres, membranes etc.) are quite similar (Fig. 4). Even the primitive ancestors of man, the unicellular ocean organisms, had similar characteristics. A surprising relic of our evolution which still resembles the ocean, is the fluid medium in which our cells reside. The circulation of blood enables the rapid transportation of oxygen, nutrients and transmitter substances as well as waste products, and it is necessary for the preservation of the fluid environment for body cells. In most cells, basic cellular functions and cell division and reproduction take place in a very similar manner, and can therefore be used for experimentation under artificial conditions (cell cultures) (Fig. 5). In the following chapters, the importance and specific use of cell cultures for experimentation and research will be discussed.

The use of cell cultures as an alternative method to animal experiments is quite promising, but are these expectations at all justified? What can we learn from a cell when searching for new substances, testing the efficacy of pharmaceuticals, or in the elimination of harmful or undesirable effects? An answer to this question is only possible, if exact measurements of cell functions in the cell culture can be performed, and if we know the limits of the applicability of these results to a particular organ or the entire human organism.

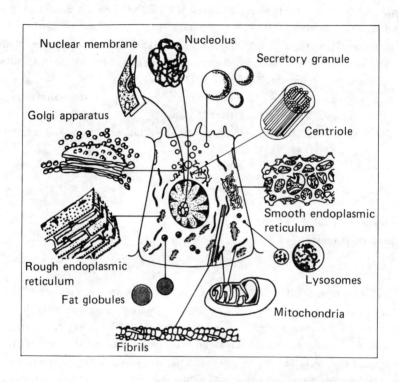

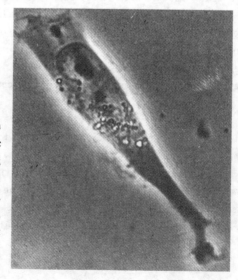

Figure 4. Cellular organisation and shape. (a) Diagrammatic structure of the cells showing some important structural elements (from: T.H.Schiebler, U.Peiper; *Histologie*, Springer, Berlin 1984). (b) Light microscope picture of an embryonal myocardial cell (phase contrast).

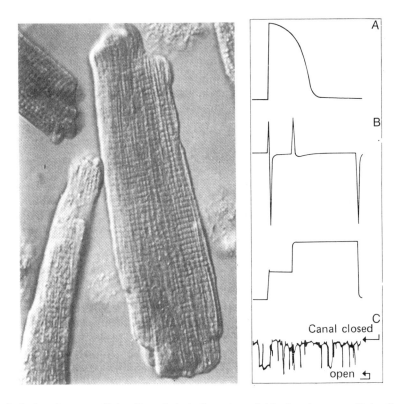

Figure 5. Isolated myocardial cells and their functions. (a) Isolated myocardial cells in cell culture (Nomarsky interference contrast). Electrical activity and contraction can be measured simultaneously in such cells. (b) Electrical activity pattern in the cell membrane (action potential, A). Measurement of electric potential across the cell membrane (sodium and calcium channels, B), opening and closing of individual ion channels where ions can only enter or leave the cells in the open position (C).

However, the moral and ethical justification of cell cultures must also be considered. Are cell cultures really an attractive alternative for the emotionally engaged opposition against animal experiments? In our enthusiasm over the alternative of cell culture, we often overlook the fact that the cell is the beginning of human life and carrier of the entire genetic information. Cell cultures cannot therefore be seen as alternative systems -

although free of pain and suffering - which are beyond moral or ethical restrictions. Relative ethics to protect life are not possible (see footnote, Albert Schweizer).[1]

The following example shows how we, at least in the western world, neglect such matters in everyday life. One who eats an egg for breakfast, and at the same time finds reports about experimentation with embryos scandalous, obviously has not considered the possibility that he or she may just have performed a deadly heat test on an embryo. Our actions and arguments against animal experiments are obviously still subconciously influenced by thousand or possibly million year-old taboos, in which the connection between 'useful' and domestic animals were established and not determined by profound ethical criteria. Insects, frogs, snakes and even rats seem to be a confirmation in this case, since these animals can expect little or no protection from engaged animal experiment opposers. As history has shown, the so-called animal lover is not necessarily also a friend of man, and is therefore not qualified for higher moral and ethical demands.

Similar demands played a very important role in the development of medicine. Much time passed until the scientific inquisitiveness of doctors conquered such taboos as the opening and the examination of the dead human body. This unfortunately, created further fear and panic among the population which was then displayed in horror pictures (Dr. Eisenbart and Frankenstein), falsely depicting the brutality of practicing scientists in the field of medicine. The cell however, is not accessible to the naked eye, and is not burdened with taboos and fears from man's past history. The cell is not a part of everyday experiences, is not an object one can stroke and caress, and therefore seems to be the ideal alternative.

Cell cultures also offer a number of very important advantages from the medical and scientific points of view, since they allow the elimination or reduction of many animal experiments. A large number of scientific journals entitled 'Cellular and Molecular...' attempt to verify the superior role of analysis on the cellular and molecular level. However, a complete substitute for animal experiments can never be achieved even with great progress in the field of cell culture. A part is not the whole. Even if we know exactly how one single stone will roll down a slope, we are still unable to predict the behaviour of a large number of stones, for example during a stone avalanche, whereby little interaction between the stones is possible. However, in the human body, numerous interactions between the many different cells in one particular organ or in the entire body,

---

[1] Based on Roman rights, later the theology of Thomas von Aquin and mainly on the "Ethics of Brotherhood", Albert Schweizer said: "The ethics of respect for the uniqueness of all life, does not acknowledge relative ethics. Its only aim must be to preserve and promote life. Every act of destroying or harming life under whatever circumstances it may occur, is to be condemned and is to be considered evil."

are possible. Each cell culture is actually an artificial simplification, dependent upon the selection of cells (the lung for example consists of more than 40 different cell types), the chosen environment and nutrient, as well as additives to the culture medium. Even life-threatening influences, systematic damage of the central nervous system, of the hormone control system or of the circulatory system cannot be detected.

The complexity of the origin of diseases and of possible points where treatment may be successful, as well as the variety of possible unwanted side effects (toxicology), pose a further fundamental problem. Only systems that are complex enough allow the elimination of undesired side effects, and systems that are simple enough can reveal effects and modes of activity. The following list shows the progression from the most complex system to the most simple:

- man
- other vertebrates
- lower animal forms
- isolated organ
- organ culture
- cell culture
- intracellular systems (intact cell parts)
- multi-enzyme complexes
- enzyme or receptor systems

Even though pathological changes can theoretically take place on the cellular level, as in the case of cancer, a simple cell test is inadequate. This is very important in the search for new substances to inhibit the growth of cancer, and to exclude the carcinogenic or genetic effects caused by new pharmaceuticals.

The Ames test on bacteria (mouse typhus bacteria) is the most well known test used to prove carcinogenic or genetically-damaging effects. These one-celled organisms usually produce the amino acid histidine, necessary for growth and reproduction, themselves. Due to a genetic defect, some strains of bacteria must obtain histidine from an outside source. If this indispensible protein-building amino acid is not contained in the culture medium, the colonies of bacteria stop growing (this can be observed with the unaided eye). This genetic defect can be eliminated to a certain degree by reversed mutation. Those bacteria which become able to produce histidine keep reproducing. If test substances raise the frequency of re-mutation due to their influence on genetic material, carcinogenic or genetic defects are to be expected. Approximately 70-80% of cell-tested substances can be properly classified with this simple test. A missing defect however, does not guarantee that genetically-damaging or carcinogenic effects are non-existent. A long term experiment with rats, in order to eliminate carcinogenic effects, takes at least two years and costs approximately 1.5 million DM. The Ames test can be completed in about two days and costs around 300 DM. The simple bacteria, cell and tissue cultures are the first choice, since they save time and money. Their evidence of harmful effects, or absence of desired effects, reduces the number of animal experiments, by filtering out those substances which are most likely to carry risks or have unwanted effects.

At present, intensive work is being done, particularly in the field of toxicology, to improve the information obtained by these tests. It is possible to render carcinogenic or mutagenic substances ineffective by catabolic removal in the body, mainly in the liver, or they can be newly formed by metabolic processes. This 'digestive process' can also be simulated by adding metabolically-active liver cell components to the Ames test. Unlike the genetic material (chromosomes) of a cell - from unicellular organisms to man - the genetic material of bacteria is not surrounded by a nuclear membrane and is arranged as a single string or in ring form. For this reason, carcinogenic substances which influence the nuclear membrane remain undetected in the Ames test, and many scientists therefore work directly with animal cells.

These cell cultures play an important role in demonstrating another type of damage which does not directly affect the molecules of the genetic substances, but damages the self-healing systems of the cell. Not every cell, whose genetic material has been damages by a substance or by penetrating radiation (e.g. X-rays or radioactive sources) becomes a cancerous cell. Repair systems and growth-regulating substances in the cell can prevent this to some extent. This self control mechanism can be inhibited by cancer-promoting substances (tumour promoters) and allows cells with damaged genetic material to be transformed into cancerous cells (cell transformation). This phase of cancer development can also be investigated in detail in cell cultures. The main problem posed by such investigations especially with cell cultures when searching for new cancer-inhibiting agents, lies in the applicability of these results to man. Even when the starting material for the cell cultures is obtained from operatively removed tumours from man, extreme differences in the sensitivity to anti-cancer substances can be found among different patients having tumours of the same kind. Beyond this, some substances may be ineffective on certain tumours from animal cells, but quite effective in humans with the same or similar cancer types. The validity of results of cell cultures and their applicability to man becomes more problematic when many different cells or organs are affected, such as in the case of the most common cause of death, heart and circulatory disease.

Which cell cultures are primarily used today?
1. *Primary cultures*: Freshly isolated cells which are obtained from organs of killed animals and are preserved in a nutrient solution (culture medium). Those cells are

generally very short-lived, lose their organ-specific properties, rarely divide and their function largely depends on the special composition of the culture medium. Primary cultures from embryonic tissue depend less upon the culture medium and divide more rapidly; the cell functions however, are at a more primitive level. If not only one cell type of an organ but more cells, often in an organotypic cell assembly, are cultivated, we speak of tissue or organ cultures. In addition to the change of organ-specific functions, the composition of cell types may change quite rapidly, due to the different speeds of growth and multiplication.

2. *Permanent cell lines*: The cells of these strains live for an almost unlimited time in a suitable culture medium, but lose their organ-specific functions quickly and completely. At the present time, the ingedients required for a nutrient solution which would ensure the continuation of organ specific functions, is still unknown.

3. *Tumour cell lines*: Tumour cells in culture are practically immortal and exist as cell lines for example from long-deceased patients. Tumour cell lines from specific organs (e.g. the liver) have gained in significance and can find unlimited use as permanent cell cultures, since they do not lose their organ-specific functions (e.g. liver metabolism or enzyme content). Specific cell culture combinations such as heart muscle cells with cancer cells exist, and are used to investigate the speed of penetration of tumour tissue into normal tissue.

The development of vaccines and establishment of their quality and safety which formerly required innumerable animal experiments, can today be performed in cell and tissue cultures to a large extent, e.g. tetanus, polio, yellow fever or rabies vaccines. Cell cultures also gain increasingly in importance in the rapidly developing field of immune biology in which some of the tests, e.g. the investigation of the compatibility of donor and recipient tissues by means of white blood cells in a cell culture, may also gain in clinical importance. The production of interferon, a substance which has a different composition for each species and triggers the body's intrinsic immune system against viral infections, and may possibly activate it to fight cancer cells, is also based on cell and tissue cultures.

The fact that organic changes caused by disease which are visible with the naked eye are the results of characteristic changes in the respective cells of these organs, led to the insight that most pathological processes occur on a cellular level. The pathologist can determine the actual pathological process and the characteristic consequences for the cell structure, only with the aid of a light microscope and the microscopic techniques of tissue preparation. The victory of cellular pathology has already made history. The measurement of functional changes in living cells is incomparably more difficult than the detection of structural changes in fixed (dead) tissue, although these techniques have been significantly refined in the past decade. Cell physiology (normal cell functions), cell pharmacology (the effect of active substances on normal or disturbed cell function), as well as cell pathophysiology (alteration of cell functions by diseases) make rapid advances due to the progress of these methods, and therefore the significance of results obtained from cell cultures is greatly improved. Heart muscle cells for example, show that measurements on the cellular level are possible and provide more information than experiments on parts of the heart *in vitro* or on the heart as a whole. Fig. 5 shows an isolated heart muscle cell (guinea pig) and different cell functions that can be measured on a single cell. In addition to the genuine measurements of the electrical activity of the cell, the accurate measurement of the ion current through the cell membrane is possible. Even the opening and closing of single tunnel proteins in the cell membrane, a rapid molecular process, can be determined directly (see Fig. 5). Since substances can also be introduced into the cell which cannot normally enter, and the entire inner content of the cell can be varied in its composition, it is possible to investigate intracellular and extracellular sites of action of substances. Spectrophotometric methods, and the use of newly developed chemical compounds which make molecular changes visible, allow the determination of rapid processes within the cell such as the variation of activity of Ca^{++} ions during a heart cycle. Accurate measurement of the variation of contracting filaments during a cell contraction (sarcomer lengths) is possible with the use of a laser. A wide spectrum of modern biological, biochemical and biophysical methods is available which can be applied to cell cultures and therefore enhance the significance of the experimental object 'cell'.

Due to the great progress and success in cell biology, cell cultures have a promising future in medical and pharmaceutical research, and will reduce the number of animal experiments (and save substantial costs) in many areas.

Clinicians turn to rats
From clinical observation to animal experiments
F. Lembeck

In a technical age the development of a new drug is regarded rather like the construction of a new type of ship. The requirements for shape, size, fittings and price of the new ship must all be established. After this 'brain storming' the construction plan is produced and the ship's keel is laid. After its launch it is seen to float and a start is made on the fittings. The captain boards with the crew and the ship puts to sea. In contrast, manufacturing a new drug is now actually more expensive than building a luxury liner. The 'building time' is considerably longer, and only after several major clinical trials is it known whether it is actually therapeutically capable and safe.

It is characteristic of drugs that with both newly developed and established agents, new questions of clinical safety arise, which can only be answered in the laboratory, often only by animal experiments. Such questions may present because of new forms of administration, clinically observed side effects or as a result of altered dosage. As illustrated by the following examples, corresponding research therefore takes place continually in the laboratory.

From Hippocrates to prostaglandins
Aspirin (acetylsalicylic acid) is not only one of the most widely used drugs, but also one of the oldest. As early as Hippocrates, physicians used preparations made from willow bark which contained a glycoside from which salicylate may be isolated. It was probably the temperature-lowering effect in infectious illnesses, and the anti-inflammatory action in various joint disorders by which these willow bark preparations mainly demonstrated their beneficial effect. Salicylic acid was first isolated and synthesised in the middle of the last century. This was fortunate since so much willow bark was then required for the manufacture of the extract that the basket-making industry was hard pressed to obtain raw material. The synthetic acetyl derivative of salicylic acid, namely aspirin, is better tolerated than its salts, but becomes salicylate again after its resorption. High doses of salicylate were the only treatment of acute rheumatic fevers for generations, while medium doses were generally used to lower temperatures (unlike the present situation, most hospital patients had fever at that time!). Occasionally, small doses were effective against headaches, a common problem today. It was only after the discovery of prostaglandins (Euler, 1932), that Vane (1971) demonstrated the mechanism of action of salicylates. Animal experiments showed that the anti-inflammatory, temperature-lowering and analgesic actions of aspirin are based on the same mechanism, namely the inhibition of prostaglandin synthesis. This knowledge opened up another therapeutic application; certain prostaglandins promote the aggregation of blood platelets, the first stage in blood coagulation. Blood coagulation is reduced by aspirin through inhibition of prostaglandins, and thus long-term low-risk prophylaxis for thrombosis can be achieved. Reverting to an old medicine in the laboratory, and to animal experimentation, thus brought not only the explanation of its therapeutic mode of action, but an additional, therapeutically important application.

Only a balloon probe
Ergot extracts exert a labour-intensifying action although it is not known how this effect was discovered. A few decades ago the isolation of alkaloids from ergot was achieved. One of these alkaloids, namely ergotamine, contracted the uterus and a purified form was subsequently used in midwifery. However, midwives were not convinced of its advantages. Finally the English gynaecologist Chassar Moir compared the effect of the purified ergotamine with the previously used extract. He introduced a balloon probe into the uterus

of women in labour, similar to that performed in the experimental measurement of intestinal contractions in the dog and which has recently been miniaturised to open up sclerotic coronary arteries. Back to the laboratory! Here chemists and pharmacologists discovered a further alkaloid, ergobasine, with an exclusive and very marked action on the uterus. From this a semi-synthetic derivative with an even better action was developed, which is now used by itself and which is free of any effects of other ergot alkaloids (Berde and Schild, 1978) (Figure 6).

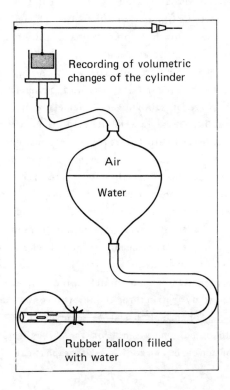

Figure 6. *Animal experimental methods and their clinical use*: The balloon probe portrayed was used in animal research to measure intestinal contraction. A small rubber balloon was drawn over the end of a glass tube and filled with water. The balloon probe was introduced into the gut. Water was forced into the container of the apparatus by the intestinal contraction, pressing air into a cylinder the plunger of which rose, so that the pointer at the top end records the contraction. The English gynaecologist Chassar Moir (1935) introduced this simple apparatus into the uterus of a woman in labour to record her contractions. Through these experiments, uterine contraction due to ergo was discovered: a drug whose action still remains unsurpassed.

Specialised, small balloon probes may now be introduced into blood vessels. This example is used for many methodological developments, which form the basis of fully developed instruments for clinical diagnosis and therapy. Today, uterine movement and fetal heart sounds, without the presence of a balloon probe, can be monitored throughout the whole course of the birth (from Chassar Moir, *Proc. Roy. Soc. Med.* 28, 1660, 1935).

A private patient
A Dutch merchant visited the Viennese specialist Wenckebach in 1912 because of frequent episodes of irregular pulse. Dr. Wenckebach confirmed vestibular arrhythmias but could offer no treatment. The patient retorted that he knew how to treat himself, and the next day had regained normal heart rhythm. During a stay in the colonies he observed that the attacks subsided when he had took quinine to prevent malaria. This observation triggered numerous animal experiments into the mechanism of arrhythmias, which were first carried out on dogs *in situ*, and later on the isolated vestibule of the rabbit heart. Today, such studies may be performed on the multinucleated cells of the embryonal myocardium. This constantly widening animal experiment potential gave an adequate insight into the pathophysiology (see p. 86). What began as an observation on a patient, led, through stages of animal experimental progress, to foundations which have facilitated the development of modern anti-arrhythmic agents.

Angina pectoris
Nitrate compounds are also the agents of choice today in the treatment of angina attacks. Brunton discovered this effect in 1867. He was caring for a patient with very frequent attacks of angina. He repeatedly checked the patient's pulse and thus established that the pulse was 'firm but slight' at the start of the painful attacks, which is a sign of a severe narrowing of the peripheral arterioles. Through the experience of chemists it was appreciated that inhalation of amyl nitrite leads to immediate reddening of the face and to a fall in blood pressure, due to marked widening of the vessels. Brunton suspected that a sudden peripheral vascular narrowing leads to a great strain on the heart, which triggers the angina attack. He therefore administered amyl nitrite by inhalation at the next anginal attack - with immediate success. Animal experiments over several decades into the physiology of cardiac contraction offered an alternative explanation. The painful angina attack is triggered by an inadequate supply of oxygen to the myocardium, which is caused by a disease of the coronary arteries. The vascular narrowing correctly observed by Brunton is only a result of the pain occasioned by the attack. It was then realised that nitrites widen the coronary arteries similar to their effect in skin vessels, although the diseased coronary arteries could barely be further widened. It was first appreciated, through further animal experiments and clinical investigations, that nitrates predominantly dilate the veins, reducing the work of the heart and allowing the myocardium to recover from its lack of oxygen. It is clear that Brunton's correct clinical observation was combined with a false assumption. Only by animal experiments and finally clinical studies could the complicated circulatory reflexes leading to angina attacks be explained. Only thus was it possible to lay the foundation of modern therapeutic possibilities.

Hidden genetics
In 1955, a synthetic curare-like muscle relaxant, namely suxamethionium, came into clinical use. Unlike curare it had the advantage of a shorter action, and the degree of muscle relaxation was dose-related, avoiding the danger of respiratory paralysis. However, one surgeon reported a dramatic suxamethonium-related incident in which a patient developed respiratory paralysis lasting several hours. He thought that the dose must have been too big, and sought to check it immediately. The action was checked on an anaesthetized cat, and the usual short action of suxamethonium was confirmed. There was no respiratory arrest, and the dosage given was correct. Undoubtedly the surgeon mistrusted the results from the animal experiment!

Two years later it was discovered that some individuals lack the enzyme which rapidly metabolises suxamethonium in the body. This is a genetic defect which does not cause adverse effects and which remains unknown - until this person is given suxamethonium before surgery. Until the enzymatic breakdown of suxamethonium was discovered in animal experiments, it was not possible to explain the cause of these rare incidents. However, it is now possible to compensate for this genetic defect by infusing the missing enzyme before suxamethonium administration.

A hormone as a poison
Certain tumours are not dangerous because of their rapid growth and spread, but because they produce a particular hormone in enormous quantities, release them into the blood and thus poison the body. Some years ago a clinician told of a 14-year old girl in whom he suspected a noradrenaline-producing tumour of the adrenal gland of being the cause of her severe hypertension. Using a relatively sensitive and specific method for detecting this substance in anaesthetized cats, high amounts of noradrenaline in the blood sample - an indication for surgery - were confirmed. The tumour was removed and shown to contain 50 times the lethal amount of noradrenaline. The child was cured by its removal. Ten years later the rat was used for the biological detection of noradrenaline. Further development finally led to a very sensitive and specific physical method (see p. 105). Today, the quantitative detection of such hormones in blood or urine is possible on suspicion of a hormone-producing tumour and this leads to diagnosis. However, without the earlier animal experiments, neither the discovery of this hormone, nor modern clinical methods of determination would have been developed.

Mental relaxation
A final example should demonstrate the extent to which a clinical discussion can lead to an unusual discovery. Penicillin is a very sensitive compound which is unstable in aqueous solution. To start with one must fill an ampoule with the sterile, dried sodium salt of penicillin G, which should be dissolved in water shortly before the injection. This results

in the desired 'injection-ready' penicillin solution. Berger discovered a particular solvent in which penicillin was stable, although its pharmacological and biological effects were unknown. With the first injection of the solvent into rats Berger had a great surprise. The animals displayed such a severe loss of muscle tone that they could hardly move, but breathing remained perfectly stable. Thus, the substance could not be used as the solvent for penicillin, but the unusual muscle relaxation was very interesting. Animal experiments showed that this substance inhibited motor reflexes. Chemical analogues showed a even stronger effect which was without toxic side effects at high doses (see 'Screening', p. 26). It was then possible to use the most effective agent of the series for the clinical treatment of muscle spasms and to discover the next big surprise. The drug produced both muscle and psychological relaxation; 'an anti-anxiety drug', the first modern tranquilliser was discovered. Was it not the researcher's duty to investigate such an unexpected observation in experimental animals?

References

Berde, B. and Schild, H.O. Alkaloids and related compounds. Handbook of Experimental Pharmacology. Vol. 49. Springer, Berlin 1978.

Vane, J.R. Inhibition of prostaglandin synthesis as a mechanism of action for aspirin-like drugs. Nature 64 (1971) 495-502.

Pointless repetitions?
The prevalence of information and animal research
W.H. Hopff

The problem of nomenclature
It is a human peculiarity to strive for individuality, in spite of, or directly because of, population density. It is therefore understandable that amongst different populations, there are differences of opinion. These primarily concern religion and politics if a particular group of people is studied, but variable opinions often prevail also in the 'cells' of the country, i.e. in families. Unfortunately, only in rare cases is there any attempt to resolve such differences in opinion using the highest achievement of the human 'animal'; understanding. Here we should not forget that overt 'animal' emotion governs our actions, where understanding is quite clearly turned off. However we must come to terms with the fact that the differences between 'informed' and 'uninformed' people, and also between 'informed experts' are now considerable. The sciences themselves proceed in a Tower of Babel. It is therefore no longer possible for the scientific specialist to understand the basis of the technical fruits which grow out of his specialty. For example, what does the average chemist know of the thermodynamics of a moon rocket engine? Or what does the physicist who builds telescopes know of the technical background to a nuclear reactor?

But how much greater is the difference between 'the man in the street' and a specialist of any kind? Here we are not just considering problems of nomenclature which can lead to differences in meaning, but genuine educational problems.

As to nomenclature, 'animal protectionists' are always referred to here as 'opponents of animal experiments', since many animal experimenters are also animal protectionists who have done much for the wellbeing of animals. All animal protectionists who reject animal experiments egotistically place themselves on an 'ethical' platform on which they force those researchers who perform animal experiments into the role of animal opponents. They malign animal experimenters and insinuate that their work is founded on evil and sadism. This does not correspond to the facts, since they who perform animal experiments feel an obligation not just to mankind, but also to animals (see also pp. 184-194). One should also keep the matter in perspective. Compared with the therapeutic uses and benefits to thousands of individuals, *both man and animal*, the number of experimental animals which are sacrificed for disease research, pure research and drug testing is negligible. If one compares the amount of animal protein from slaughtered animals to provide food for mankind, with the amount from experimental animals, the difference is even clearer. However, the benefits which are obtained through knowledge acquired from animal experimentation is significantly greater for mankind.

Information flood against intellectual training
How is it possible with the current information explosion that the discrepancy between the 'uninformed' and the 'informed' is continually increasing? The answer to this question can be given by anyone who gives it some thought. Actually, the first question induces a second: who gives out the information? With the answer to the second question it would be impossible, even for a very tolerant critic, to restrict himself to a factual argument, since the realities have become too far removed from objectivity. As an example, the animal protectionist initiative of one Herr Franz Weber in Switzerland may be useful.[2]

Rejection of this initiative indeed needed the votes of a critical electorate. Considering the low number of the voters, with almost all the protagonists going to the ballot box and yet suffering such a heavy defeat, one must believe in a 'sound national feeling'. The damage done to Herr Weber and his 'animal protectionist' friends can hardly be measured. Herr Weber had not obtained any improvement in animal protection, nor could he understand the simplest biological implications. Neither could he grasp the point of university research or even the need for animal experiments in a responsible

---

[2]The "national initiative for the abolition of vivisection" of Herr Franz Weber from Montreux was rejected on 1st December 1985 by 80% of the Swiss. The proposal of the national initiative would have effected an immediate suspension of all animal experiments.

pharmaceutical industry. Nevertheless, this attitude has achieved respectability in certain circles through the media support for Herr Weber, a respectability which cannot be reconciled with its credibility and competence. In addition to this classic example, any television viewer has seen how people talk on daily so-called 'chat shows' on a theme which neither the presenter nor any of the discussants really understand. The outcome of such a discussion is often accepted by the uninformed viewer. Here it is interesting to note, that in the eyes of the general public, film stars are endowed with the highest degree of credibility and, unfortunately, competence. A competence of similar importance is also attributed to journalists. Can it therefore be surprising that even the 'informed' sometimes believe all they read in the papers?

When errors are discovered, and it becomes public knowledge that the media have disseminated false information, the right of an expert who wishes to comment and correct these errors is often suppressed. Many scientists who have dealt with the media can tell a similar tale. If one examines the basis of many differences of opinion, one quickly finds communication problems, which are solely due to nomenclature. Even a shortage of information plays a significant role.

There are several great problems of informing people: How can information be disseminated in accordance with the truth? Who is responsible? Do purveyors of the media, and journalists working in the media, have the specialist knowledge and all the information needed to be able to assess a subject field critically?

Lack of information
Two examples should show that lack of information leads to erroneous conclusions, and indeed that assertions are made which, on closer inspection, cannot be sustained.

Thus the animal protectionists hold Christian beliefs and generally place 'Christianity' to the forefront. The Christian attitude forbids any animal research! They seem to ignore that Christian institutions have quite thoroughly investigated the problems of animal research. A remarkable example is the stance taken by the 'Assertions on the Relationship of Man and Animal' of the Institute for Social Ethics of the Swiss Evangelical Church Society. With an eye on the voting of the 'Weber Initiative' to discontinue animal research, the Evangelical Church Society was impelled to publish the following declarations:

1. Animal experiments must be well founded and even then are only permissible if there is a higher gain than the injury to the animal (justification criterium).

2. Moreover it must be seen that the appropriate animal experiment is effectively indispensable for attaining this goal (usefulness-limited criterium).

3. A continuing reduction in animal experimentation must be achieved (minimisation criterium).

4. The preservation of these criteria need strict and active controls (control criterium).

The Institute for Social Ethics (ISE) of the Swiss Evangelical Church commented on this as follows :

The basic ethical understanding of the ISE is the dialectic (encompassing opposites). On the one hand mankind as the crown of creation, and as the lord of Nature has power of disposal over the animal and he can use it for his own uses. On the other hand, the animal has an individual value which man must respect. Man and animal are fellow creatures. The use of and respect for the animal stand in indissoluble relation to each other.

As a directive in dealing with the legal utilisation of animals the ISE adheres to this. Neither is the animal merely an agent for human self-expansion, nor can it claim unlimited protection. The ISE pleads for the legalised use of animals which takes a serious attitude to the balance between respect for the animal and the use made of it by man.

From these fundamentals the ISE agreed to reject the initiative because of its demand for the absolute priority of animal protection. How do researchers who perform animal experiments stand in relation to this?

At the outset it was indicated that those researchers who perform animal experiments were forced by animal protectionists into a position of being 'animal opponents'. But what is the position of animal protectionists who themselves carry out animal experiments? These individuals have always reconciled their profession with the declarations of the ISE.

Certainly the doctor who carries out animal experiments may propound a further criterium: man should, despite progress, recognise with humility that numerous problems of medical science remain unresolved, and that there is a duty towards *man and animal* to further pursue purposeful research (duty criterium). This obligation extends not only to the research clinician, but also to individuals engaged in pharmaceutical research in the so-called 'industrialised nations'. This requires an appropriate background of collective experience as well as dedicated research personnel, research institutions and financial resources. It is thus also a duty to promote research for those developing countries which cannot realise such research for themselves, but whose people suffer from poorly researched diseases. The general public may know of the diseases which are being studied in various research laboratories, albeit not by name. Who amongst us still thinks of leprosy even though many of the poorest leprosy sufferers die in developing countries in dreadful conditions? In addition to leprosy there is also malaria, sleeping sickness, leishmaniasis, ankylostomiasis, bilharzia and many more infectious diseases which must be studied in order to develop the appropriate remedy. For this purpose animal experiments are absolutely vital. In this respect, it need not be stressed that both domestic and wild animals are also afflicted by these diseases. Thus, the remedies developed by the pharmaceutical industry are not just used for domestic animals, but also on those in the wild, in order to cure them and save them from extinction.

However, this obligation has a further, no less important significance which concerns psychology in general and that of youth in particular. In certain decadent circles the so-called 'no future' philosophy prevails. But a young person who believes in a future must be orientated towards progress, that is to say he must be research-oriented; after all our forebears did not sit twiddling their thumbs.

As a further example the efforts of industry and the pharmaceutical industry in particular can be cited.

The pharmaceutical industry is seen by animal protectionists as purely profit-orientated. They insinuate that numerous research animals are slaughtered for the sole purpose of making money. Anyone who appreciates the basic attitude of private industry and thinks that only a well organised business is profitable, comes to another conclusion: why should one carry out very expensive animal experiments if simple *'in vitro'* methods could be used for the same goal?

The life work of Arthur Stoll
Since the time of William Withering it has been known that the foxglove contains substances which are beneficial in the treatment of heart disorders in older people. However, Withering knew that in treatment using plants or their endogenous substances only one thing is certain; the uncertainty of the constituents of the active substance! Should the active agents be tested directly on the patient whose vital functions were already considerably disordered by his cardiac disease? No, this is where animal experiments came in. However, since animals generally have healthy hearts, the question of how to test the cardiac agents was raised. Testing was done for toxicity and the amount of crude foxglove extract required for cardiac arrest was determined. The guinea-pig proved to be the most suitable animal, and provided a reference point for therapy. Thus, severe toxic effects, as well as equally dangerous underdosage, could be avoided. On the other hand, this testing cost the lives of thousands of guinea pigs. The optimal dose had to be individually determined for each patient. Every new batch of foxglove extract

presented the doctor in charge with new problems. Finding the optimum dose was also a vital problem for the patient, and often involved damage to his health.

Arthur Stoll spent a period of his life working out complicated chemical structures. At a high point in his research he succeeded not only in determining the purity of the active agent, but also in clarifying its chemical structure and action. Since then, the pharmaceutical industry, in this example the firm of Sandoz in Basel, has been able to produce pure, well-defined chemical compounds of consistently high quality. Once the doctor has established the best dose for his patient, he can rely on it and be confident of always getting the same preparation again. Thus, he can continue to provide reliable therapy, but without the need for numerous animal experiments.

The accusation of repeated animal experiments
Animal researchers are often accused by animal protectionist circles of conducting 'unnecessary' animal experiments. For example, it may be that a team in Canada would be doing the same animal research as a group of Indian workers.

There is a valid objection to this argument; animal protectionists who support it are ill-informed. This is not a specific dearth of information, since an animal protectionist, who completely rejects animal experiments, would certainly have ensured that he was fully informed. However, this information mainly originates from animal protectionists and is therefore biased. In contrast, most animal experimenters are well aware of the animal protectionist literature, and are forced into an completely defensive position by these individuals. For this reason it is vital that animal researchers are adequately briefed on the arguments of animal protectionists. On the other hand, the latter are generally good hearted people who believe in their 'mission', but who have little factual information or are even not interested in being informed. This would lead to reconciliation of differing perceptions, which is undesirable to certain animal protectionists. Is not the rejection of five different animal experiment initiatives in Switzerland, starting in the last century, the best evidence that the average citizen was better informed on both points of view than the animal protectionist?

This is a reference to the national votes in Zurich (1895), Bern (1903), Zurich (1924), Basel (1939) and the Swiss Federal vote (1985). Only someone who is familiar with the history of medicine can imagine how things would have been for us today if one of the earlier initiatives had been approved.

There are at present various ways of disseminating information. Apart from newspapers, books and magazines and radio, the spectrum of the 'media' has now broadened considerably.

The assertion that the scientist would make too little effort in publishing new results can be decisively countered, as the following examples show:

1. Giving information.

Today's potentials for preparing and spreading information are almost unlimited. Microelectronics and computer science have made a crucial contribution to this.

a) Conventional ways

Ever since man learned to write, the literature has expanded. The art of writing experienced its first impetus when printing was invented. Compared to the time it used to take to disseminate written information with the time needed today, a tremendous acceleration has been achieved. A smooth transition is currently being made from conventional printing methods to electronic text processing. *Current Contents* is a good example of this. It puts almost the entire world's research at our disposal. Within a few weeks of its appearance, the title of an article and the authors' addresses are listed in this special publication. Thus, even if the paper required cannot be found in any of the innumerable university libraries, one can write directly to the author and obtain a reprint of the work.

In addition to this information, numerous other publications are available to researchers. Specialist journals may be found in every modern institute. None of these scientific journals will accept a paper for publication which only repeats previous research without any new findings. Furthermore, specific reports on progress are available to industrial researchers and the appropriate bulletins are at the disposal of university scientists.

All this has not been mainly due to pressure from animal protectionists however, but has been the case for longer. One of the earliest information bulletins is a good example. University scientists interested in neurobiology have, for over 16 years, been publishing an information bulletin called *Neurobiologie Zurich*. In the introduction to one issue, Professor W. Lichtensteiger wrote:

> "This year's bulletin goes out following the vote on the national initiative 'for the abolition of vivisection' of Franz Weber. The clear rejection of this initiative by the nation is of the greatest importance for neurobiology. In the run-up to the election there was often talk of a lack of mutual information between scientists. The Zurcher Neurobiology Bulletin has been appearing since 1972. Its existence shows, that to those directly involved, the importance of reciprocal information exchange was realised at a time when this still did not constitute a general talking point. Information for the

non-specialist poses another problem. For him the bulletin in its necessarily concise form is unfortunately difficult to read. However, the discussion about animal research has again shown the importance of continuing factual information for the public in understanding the significance of pure biomedical research. Further efforts in the future will be needed in this area".

Without the intensive communication which is practised today in all universities and research centres, modern research would not be possible. It does not need to be stressed that such rapid communication largely eliminates any duplication of research.

b) Electronic media

The telephone may well be considered the first of the electronic methods. It is used not merely for verbal communication, but allows access to most of the world's databanks. The databanks themselves may be connected to each other so that almost all the data one could wish for is readily available. However, databanks are not only used for literature searches; they can also be used for complicated calculations and structural analyses. The connection of three important American computing centres is a classic example. Professor Edgar Meyer of the Texas A- and M-University designed and constructed complicated enzyme and receptor models, which are now used as the basis for research without using experimental animals. Certainly, the use of research animals can be bypassed with these methods, although the resultant data must then be tested on living organisms.

c) Computer simulations

The computer already has a secure place in education. Taking experimental data as a basis, computer simulations may be used for teaching purposes. Certainly these can only replace research animals to a limited extent, and one must realise the limits of this approach (see p. 106). Thus, cooperation between animal protectionists and animal experimenters can turn out to be a good thing. Would it not be better for the animals if the animal protectionists and the animal experimenters sat down around a table to do justice to the wellbeing of the animal? Such collaboration was seen in the 'Hildegard-Doerenkamp and Gerhard-Zbinden Foundation for Realistic Animal Protection'. Frau Doerenkamp as the true animal protectress, and Professor Zbinden, the Director of Zurich University's Institute for Toxicology, gave the impetus for worldwide cooperation. Contributions with worthwhile programmes for saving animals came in from everywhere; from Russia and Central Europe to New Zealand. The submissions were evaluated by the following criteria:

- Usefulness in teaching
- Saving animals
- Originality

— Scientific level

It was not just for the sake of awarding a prize to the best work in order to give further stimulus to researchers, but also for the world-wide publication of the work which did not achieve first place. With currently-available audiovisual and computer methods, animal experiments for educational purposes can be waived in many medical disciplines. Furthermore, with the computer, almost all media for information dissemination can be fully exploited. The level of work submitted was astonishingly high, not just in its scientific content, but also in its educational theory and programming technique.

2. Diversification
Research has now diversified to the point where, on statistical grounds, it is unlikely that the same work is being pursued in two places in the world at once.

3. International conferences and workshops
Researchers working in the same field often know each other. Contacts are made worldwide, particularly at international conferences. In addition, so-called 'workshops' - small working groups concerned with a specific subject area, permit good personal contact, and, with the worldwide financial crisis, by the pooling of problems, make savings in both money and animals.

How does it look on the other side?
On the other side, one has only to read the commentaries of the animal protectionists to realise that it is rather a case of insufficient information. These commentaries are always based on emotions, and often reach a peak in personal attacks on research personnel. Researchers are represented as sadists who inflict distress on animals for pleasure. The research goals are never mentioned.

Has giving information any disadvantages?
This question must be answered decisively in the affirmative. Indeed, it is clear that attempts are made to use the worst kind of 'journalism' to incite the ordinary citizen's lust for sensation. How could it ever be possible that such a 'neomysticism' could spread throughout the world? In medicine in particular, where alone the natural sciences have helped to bring about a breakthrough, ignorant 'experts', who hardly ever passed a medical examination, make unfounded assertions. Natural laws are thrown out or questioned with astonishing ease, and centuries-old ideas are resurrected, whose worthlessness should be clear to every secondary school pupil of average intelligence. The reappearance of homeopathy, Christian Science, iridiagnosis and other pseudomedical 'sciences' might serve as examples. If these 'sciences' are disseminated by both the press and television, they increase in popularity and a strong claim to legality would be accorded them. Cautioning

scientists are debarred from intervening and discouraged from contributing to the discussion.

Even in the campaign for the Weber initiative in Switzerland, the popularity of homeopathy, which also sees animal experiments as unnecessary, was abused by the animal protectionists. This confused the public.

Thus it is evident that, on the one hand, informing people is a sensible way of disseminating effective knowledge. Conversely however, it should not be allowed to escape attention that, manipulated by the sensation-seeking media, it can cause considerable harm.

Is toxicology more than just the LD_{50} test?
New methods in *in vitro* toxicology
E. Beubler and B. Schmid

Approximately 60 years after its development, the LD_{50} test of Professor Zbinden, the experimental toxicologist and director of programmes for the reduction of the use of animals in biomedical research, is seen as 'a ritual mass execution of animals' (Figure 7). How has this happened?

Originally, the LD_{50} test was developed for the determination of highly active pharmacological agents, such as the digitalis glycosides in foxglove leaves. The LD_{50} is defined as the dose which kills 50% of the animals treated with the test compound. From this 'all or nothing' reaction (i.e. dead or not dead) in the foxglove example, it was possible to determine the glycoside content and thus the optimum dose for therapy. The 'all or nothing' reaction is still necessary today in certain cases: The action of an anti-rabies serum for example, cannot be determined in any other way than by administering the antiserum to an infected animal. If the animal survives, the antiserum is useable; if the antiserum is too weak the animal would necessarily die - and any infected person treated with such a serum. Eventually, however, the LD_{50} test was introduced into toxicology as a standard test of the toxicity of any substance - foodstuffs and fodder, industrial chemicals and pesticides. The original purpose, the determination of the toxic action of highly active drugs, was forgotten. Thus, the test has become widely required by health agencies, and is seen as a test which must be performed by researchers.

The present LD_{50} test can be absolutely absurd. Thus in 1967, one publication reported the LD_{50} of distilled water to be 469 ± 51 ml kg^{-1}. Along with the questionability of the validity of such a result, the correct statistical determination of LD_{50} simulates 'scientific methods', but says nothing about the cause of death, which, in the distilled water example, may be the sheer volume of fluid or a change in osmolarity. The most alarming further development with the LD_{50} determination was its total automation; 'dead or not dead' was no longer observed by an educated eye, but determined automatically by light beams. This results in the total loss of reference between the person conducting the test and the living creature.

Each one quickly laid an egg
And then died.-

Figure 7. A LD_{100} test on hens? This gruesome 'experiment' of the bad boys Max and Moritz would now be (rightly) punishable according to the law of animal protection.

A general move away from the LD_{50} test still does not seem to be in sight. The authorities demand, as before, that the LD_{50} is determined for all compounds. If the test is not performed, the law requires that the substance in question is regarded as highly toxic.

As a step towards the improvement of the situation however, it is possible to substitute the **exact** determination of the LD_{50} by calculating the LD_{50} range. For example, it requires numerous animals to conclude that the LD_{50} of a compound lies between 6.3 and 8.7 mg/kg, but relatively few to deduce that it lies between 5.1 and 9.9 mg/kg; the knowledge of such a range almost always suffices. In the Federal Republic of Germany for example, the exact LD_{50} value is no longer required. This has led to a 75% reduction in the number of animals needed for this type of research.

Since the determination of the LD_{50} is always associated with the deaths of research animals, toxicologists have tried to completely replace the LD_{50} test by *in vitro* methods. For several years, various toxicology centres have attempted to use cell cultures as alternatives to the LD_{50} determination in whole animals. However, there have been a number of difficulties with this approach.

It is easy to establish the death of an animal. It is much more difficult to establish the point at which an individual cell dies; numerous different biochemical processes can be impaired simultaneously or individually by a chemical compound. Whether there is a parameter which is also relevant for the *in vivo* situation remains an open question. When toxicologists can finally offer new and certain methods which are recognised as reliable by the authorities, LD_{50} determinations will be relegated to oblivion.

In vitro systems have been able to demonstrate the reproducibility of results in different centres. In a multi-centre English study, the same cell lines were used with the same substances, and identical parameters measured. The results from the different centres were almost identical.

With all the advantages and disadvantages of the LD_{50} test and the alternative *in vitro* systems, the place of acute toxicity testing within the whole of toxicology remains debatable. In 1975, Professor Neubert wrote: 'The lethal dose is indeed the easiest to determine routinely by experiment, but often gives the least information for toxicological assessment'. Furthermore, 'For toxicological assessment, the amounts which reversibly or irreversibly damage certain organ systems or are carcinogenic, teratogenic or mutagenic, are increasingly crucial. These effects hardly need to be correlated with the acute or chronic lethal dose.' According to this, toxicology means much more than the LD_{50}, and thus the number of animal deaths justifies the enormous efforts which are being made to replace this particular test.

Other toxicological methods are also in question, and alternative methods are actively being sought for such tests. Amongst these is the Draize test, which has particularly been criticised by opponents of animal experiments (see p. 46). This test was developed some 40 years ago for testing the tolerance of skin and mucous membranes (Draize *et al.*, 1945). The law requires (at least in the USA) that every body care agent and household product, from shampoos to oven cleaners, from deodorants to fabric softeners undergo this test.

In the Draize test. the test substance is dripped into the eye of the rabbit (mouse and rat eyes are not very sensitive) and the mucous membrane reaction is then observed. Damage can range from none to very severe inflammation, and can thus be experienced by the animal as perfectly tolerable to extremely painful. The test **should** be an approximation, in which it is not the concentration leading to damage which is measured, but only that which causes no damage. Therein lies the sense of this test! In 1980, a society of humanitarian groups started a campaign against the Draize test, and thus the cosmetic industry was particularly caught in the crossfire. This industry, and the American Revlon consortium in particular, responded by giving 2.5 million DM for the development of alternative methods; the reliability of various such methods is presently being investigated worldwide. In Boston, the culture of human corneal cells was developed, which appears suitable for the preliminary testing of aggressive substances. In the chorioallantois membrane test (HET-CAM-test) the skin lying directly under the shell of an incubated hen egg, containing veins and arteries but no nerves, was used as the test organ. For the determination of inflammatory substances, the CAM test is a rapid and simple method. Inflammation is however, only one of the symptoms which can be occasioned by irritant substances. Although this method also has its limitations, it allows a certain discrimination between irritant and non-irritant substances. A further possible replacement for the Draize test is the measurement of dye (neutral red) uptake into specific cells in culture. Preliminary results with detergents correlate well with those obtained from the Draize test.

The authorities displayed a more decisive reaction for the Draize test than for the LD_{50} test. They banned the Draize test for certain aggressive substances (e.g. strongly acid or alkaline substances) and were in agreement about reducing the number of animals used. Furthermore, the use of local anaesthetics was recommended, which mitigated subjective distress (pain) without influencing the result (inflammation and ulcers).

The Draize skin test has similar problems to those of the Draize eye test. In the former, substances are tested for their irritant (i.e. producing reversible inflammation) or corrosive (i.e. causing irreversible tissue necrosis) effect. Rabbits are still being used for these tests too, although human cadaver skin can be used as an alternative.

Human skin can be stretched *in vitro* and the change of electrical skin resistance is used to measure potential damage. Using this method, it was shown that out of 44 substances which seem positive in rodents (and thus cause damage), only 23 are positive in man. Thus, human skin seems to be relatively insensitive. Further dermatotoxic substances may also be tested on human epidermal cells in culture. Cell growth and membrane permeability (porousness) can be used as parameters of damage. Rat epidermis may be used for other newly developed *in vitro* methods, in which several samples can be taken from one rat. Skin resistance is also used as the damage indicator. After some initial difficulties, false positive results have been reduced to about 5% with this method.

In vitro toxicological safety testing of drugs and industrial products is always important in the eyes of industry. On the assumption that *in vitro* tests provide relevant data, undesirable properties can be established much more quickly and with considerably smaller amounts of test material. Such tests are now frequently used to recognise potentially genotoxic properties of chemicals. If such a property is detected, this still does not necessarily mean that the same toxic effects will occur in humans when used appropriately. Nevertheless, further development of such a substance would sometimes be rejected, thereby saving numerous research animals. The potential in this field is not only extremely promising, but is also independent of legal constraints, since the information obtained from such methodology is only used to aid decision making.

However, if accurate information is required about the risk to humans, experiments on animals remain indispensable. Only these tests can ultimately provide information about uptake in the organism, whether it is transported to a critical organ and causes harm, whether it is rendered harmless by detoxification processes or whether it is rapidly excreted.

Even if a single type of cell in culture can never completely substitute for the animal, it should still be noted that combinations of cell culture tests are feasible which are relevant, efficient, sensitive and economic. However, there are considerable problems in their use, which must be resolved by further research. The development of new toxicological methods is vital to keep pace with the growing numbers of new substances with which we are continually confronted, since the legal requirements of safety testing of substances are also growing rapidly. Some toxicologists can see the day when they will no longer be able to cope. It may be that there will be an insufficient number of toxicologists and not enough animals in the whole world to satisfy these demands.

References

Draize, J. *et al*. Methods for the study of irritation and toxicity of substances applied topically to the skin and mucous membranes. J. Pharmacol. exp. Ther. 83 (1945) 377-391.

Purchase, I.F.H. and Conning, D.M. International conference on practical in vitro toxicology. Food Chemic. Toxicol. 24 (1986) 447-818.

Rowan, A.N. The test tube alternative. Sciences 21 (1981) 16-34.

From animal to computer
New physical and biochemical methods for replacing animal experiments
A. Saria

Chemical and physical systems
Long before the beginning of modern technology, efforts were made to perform experimental studies of phenomena seen in living organisms, using non-living systems (U.S. Congress, Office of Technology Assessment, 1986). An example of this is the biochemistry of the enzyme, the so-called 'biocatalysts', which are responsible for almost all chemical processes in living organisms. By 1913, various isolated enzymes had been characterised outside living organisms (Michaelis and Menten, 1913). Using new physical methods such as nuclear resonance spectroscopy, the structure and function, as well as alterations of enzyme activity for medical purposes can now be studied almost exclusively *in vitro*, i.e. in the test tube. This shows that 'alternative methods' can develop without any further help if the opportunity is given to pure research. Two modern analytical techniques which were developed out of efforts to improve existing methods, are radioimmunoassay and high pressure liquid chromatography. Both methods have replaced numerous animal experiments and will be dealt with in more detail here.

Radioimmunoassay
This highly quantitative method was originally developed by Yalow and Berson (1960) for the determination of insulin. The technique is based on the reaction of an exogenous 'antigen' with an 'antibody' produced by a living organism as a defence against that antigen (e.g. a pathogenic virus). Using biochemical methods, antibodies against the target antigen can be stimulated and isolated in experimental animals or cell cultures. The purified or chemically synthesised antigen which has been radioactively 'labelled' is also required. No description of the methodology will be given here since this is dealt with on p. 115 of this volume. This method has replaced a large number of animal experiments. There are now radioimmunoassays for almost all hormones, which used to be measured in bioassays which were often associated with pain for the animals. On the whole, most animal experiments for the quantitative determination of hormones, and for which an effective radioimmunoassay exists, are now unnecessary. However, to complete the picture it should be noted that animals are still needed for antibody production, albeit in relatively small numbers.

High pressure liquid chromatography
Another method for the quantitative determination of endogenous substances is high pressure liquid chromatography (HPLC), which is based on a completely different principle from radioimmunoassay (Engelhart, 1977). In this technique, mixtures of substances from blood or different organs, are pumped at high pressure through a steel column closely

packed with solid particles. This solid matter may consist of small silica gel pellets. Different substances adhere, on the basis of their chemical properties, more or less strongly to this solid matrix, and are therefore eluted from the column at different rates. In this way, the individual components of a mixture can be separated from each other. These then pass through a detector unit which takes advantage of specific physical or chemical properties of the test substance, e.g. fluorescence, optical density or radioactivity, to detect and measure the test substance. HPLC allows the detection of very small amounts of a compound, although radioimmunoassay has a similar sensitivity. HPLC has likewise replaced a number of biological tests, for example the measurement of hormones from the adrenal medulla and certain neurotransmitters in isolated organs of rats or rabbits (Blaschko and Muscholl, 1972). The determination of these hormones, which may, for example, be secreted by the adrenal gland under stress, is necessary for the diagnosis of certain hormone-secreting tumours.

The computer as a substitute for animal experiments
The use of computers in biomedicine has brought significant innovations which may be divided into three categories:

1. The development of completely new physical methods for diagnosis, therapy and research which would be impossible without microprocessing techniques. The examples of computed tomography and NMR spectroscopy, which make it possible to see inside the body with remarkable definition, should be mentioned here.

2. The use of computers for the recording and mathematical assessment of measurement data in research, hospitals and industry.

3. The computer simulation of biological processes.

The applications mentioned in group 1 are used directly and primarily for the patient who can thus perhaps be treated more quickly and effectively. However, these applications also contribute indirectly to a reduction in the use of animals, since problem-free studies on humans lead to knowledge which was once sought *via* animal experimentation.

The recording and mathematical handling of data mentioned in group 2 has improved measurement accuracy. Since repetition of the same experiments and measurements always leads to variable results due to biological differences in experimental animals, instability of measuring equipment and observer error, a particular experiment must be repeated several times. Thus, small differences, which may for example occur due to drug treatment, must be established by statistical methods. More accurate measurements therefore help to reduce the number of research animals needed. Furthermore, it sometimes happens that - regrettably - irresponsible researchers, through ignorance or laziness, and possibly simply because they are easier to calculate, choose statistical procedures for their data which are inappropriate. This results in conclusions which are not supported by the data, and therefore renders the animal experiment useless. Computers now make it relatively simple to perform the most complicated mathematical and statistical procedures, and should put an end to such unreliable work. Today, measuring equipment is increasingly linked to computers, and new techniques for their use are published almost daily. For example, the activity and behaviour of experimental animals, which even recently was studied tediously by subjective observation, can sometimes be recorded photo-electrically and evaluated completely objectively by computers. The development of cheaper and more powerful personal computers has resulted in such applications not being reserved merely for very costly projects.

Computer simulation is a relatively new technique which is being used to pursue research into the action of drugs, and to understand the functions of the human body. In principle, this approach attempts to reduce a bodily function and its possible regulatory mechanisms to mathematical formulae. By 'trial and error', hypotheses and animal experiments, more and more biological processes are becoming understood and converted into a mathematical form. The results from such computer programs, showing perhaps the influence of a drug on one part of the regulatory process, may then be re-checked in the animal experiment, and if there is disagreement, the program is adjusted. Agreement suggests that the computer model adequately reconstructs the biological process and thus 'simulates' it. A series of such models, including simulations of kidney, heart and lung function, individual parts of the nervous system and metabolic functions are currently under development (US Congress Office of Technology Assessment, 1986). The difficulty of this venture is illustrated by the fact that in a simulation of the heart and circulatory system for example, the complicated regulatory mechanisms of the heart, lungs, kidneys and brain, all have to be considered together. Many further mechanisms are currently being investigated with computer simulation. Among others, these include:

- the regulation and transport of sodium, potassium and calcium in the heart
- the blood coagulation system
- drug effects in the lungs
- a model for salt transport in nerves and muscles
- the function of the inner ear

Computer simulation can presently be regarded as a supplement but not a true alternative to animal experiments. In fact, numerous animal experiments are directly associated with computer simulation, in that they are required for checking and improving the computer models.

'The development of computer programs of improved usability is directly connected with the use of animals in biomedical research' (US Congress Office of Technology Assessment, 1986). Furthermore, the extrapolation of a computer model to man is only valid if the model's basic data are transferable. This means there is essentially a similar risk to that seen in animal experiments. Pratt, a scientist who clearly opposes animal experiments, dedicated a mere 12 lines to computer simulation in his 280 page book on alternative methods (Pratt, 1983). This highlights the relatively meagre place presently accorded to these methods as alternatives to animal research. If computer models should in future replace a number of animal models, some intensive collaboration would be needed between computer experts, mathematicians and bioscientists and the medical profession.

Such collaboration is certainly not yet as pronounced in Europe as in the USA. One department at Duke University in North Carolina maintains a database, initiates bioscientists in the concept of 'mathematical modelling' and provides help in the production of programs (US Congress Office of Technology Assessment 1986). The establishment of such centres in other countries would be of great significance.

References

Blaschko, H. and Muscholi, E. Catecholamines. Handbook of Experimental Pharmacology 33, Springer, Berlin 1972.

Engelhart, H. Hochdruckflüssigkeitschromatografie, 2. Aufl. Springer, Berlin 1977.

Michaelis, L. and Menton, M.L. Biochemistry 49 (1913) 333-369.

Pratt, D. Leiden vermeiden. Turm, Bietigheim 1983.

U.S. Congress, Office of Technology Assessment. Alternatives to animal use in research, testing and education. Washington DC, US Government Printing Office, OTA-BA-273, 1986.

Yalow, R.S. and Berson, S.A. J. Clin. Invest. 39 (1960) 1157-1175.

5

Special Areas of Research

Anyone who visits a modern film studio can barely imagine that a worthwhile film will emerge from the confusion. Similarly, anyone who visits a research laboratory finds it difficult to see how important new knowledge may be gained from the abundance of animal experiments and other work. In both cases there is a clear objective behind the technical and methodological complexity: In the former, a film should be acclaimed by its audience, while in the latter, a medical advance must be demonstrable.

This objective also applies to the use of experimental animals. The final aim is to alleviate human suffering. However, even in modern pain research, the suffering of research animals has long since ceased to be the only way in which human distress may be relieved. New methods, however different they may be in the various areas of research, are not 'alternative' (i.e. 'other'), but 'selective', because they provide better answers to scientific questions.

Tiny amounts - great effects
Animal experimentation in endocrinology
H. Kopera

One day in 1948, *The Times* was more interesting to many doctors than any scientific journal. It contained the news that the American rheumatologist Hench had, on 21st September, discovered an 'anti-rheumatic factor'. With a few injections he was able to remove pain and restore movement to people who had been severe rheumatic sufferers for years. This was a 'rise and take up thy bed' story which actually happened. What was the background? Hench had long suspected that the lack of an endogenous factor leads to chronic rheumatism. His attempts to isolate this factor were in vain; he suspected that it lay in the adrenal gland. Although Reichstein had already isolated some 20 steroids from the adrenals, it was still not known which of them had vital hormonal functions. Animal experiments showed that young rats survived adrenalectomy for days without apparent adverse effect. However, if they were put in a room at 4°C they died in a few minutes. Thus, they were lacking a factor from the adrenal gland which enabled them to withstand cold stress. If adrenalectomized rats were injected with a specific fraction of an adrenal extract, they subsequently withstood the cold. This test facilitated both the measurement of the important adrenal hormone in the suprarenal and in the outflowing venous blood, and also the establishment of the adrenal stimulus which triggers hormone production and output. Numerous rats were needed for this. Using such an approach, together with other animal experiments, it was subsequently possible also to isolate the active hormone - cortisone. Hench injected this compound, fortunately at a therapeutically appropriate dose. In this way, a completely new mode of therapy was initiated!

Figure 8. *Iodine deficiency goitre.* In his cosmography which appeared in 1596, the geographer Georg Munster (whose portrait appears on the 100 DM banknote), described Styria as a country in which the people were characterised by enlarged thyroid. He suspected the air or the water as the cause. In 1890, a congenital disease was cited as the cause of this thyroid abnormality. Neonates had goitre, their mental and physical development was sharply reduced and they became cretins. About every 400th inhabitant suffered from this condition.

Modern endocrinology developed on the basis of animal experiments solved the problem: the dietary iodine content was too low. Thus, insufficient iodine-containing thyroid hormones were produced. The iodination of cooking salt now guarantees, without additional cost, an adequate supply of iodine. The goitres have disappeared.

However, the adrenal glands from all the slaughter houses in the world had been unable to satisfy the demand for the new drug cortisol since adrenal glands store relatively little of the freshly synthesised hormone; they pass it immediately into the blood. The eventual synthesis of cortisone from bile acids was only achieved at great expense. Ingenuity, fortunate coincidences, and, above all, the synthesis of 'super-cortisones' with a stronger and more selective action than their natural counterparts, has since led to a rather different situation. Cortisones may now be produced in unlimited amounts, and so cheaply that they are available to even the poorest patient. Indeed, their use is so widespread that one must be wary of their overuse.

Any patient suffering from an inflammatory disease today, and who experiences relief through cortisones, may recognise from this example that the background is one of years of tedious phases of animal research. The same applies to every area of endocrine research.

Endocrinology is the study of hormones. Hormones are chemically identifiable substances (proteins, steroids, amines, unsaturated fatty acids) which serve as inter-cellular messengers. The molecule contains a genetically established message, containing information about the cells which the hormone should react with and the process which should be initiated. Hormones are active at very low concentrations. In complex ways they regulate the functions of many cells, organs and physiological systems and exert regulatory influences on growth, differentiation and metabolic processes, without themselves playing a part in energy-producing processes. Knowledge of their structure, pharmacology and the many ways in which they participate in various vital processes in healthy and diseased organisms is vitally important.

Hormones are produced in the body, either in secretory glands ('classic' hormones) or in other organs, in cell groups or from precursors in blood plasma. The former generally exert their effects far from their production site in the peripheral tissues, the latter mostly in direct proximity to their place of origin.

The history of endocrinology
Hormones are found in plants, animals and man. Their existence is most obviously expressed by the symptoms caused by their shortage or over-production. However, such diseases are relatively rare in humans in a very noticeable form. This is why animal experiments were responsible for the discovery of hormones, the elucidation of their structure and function, and subsequently, for the production of therapeutically useful hormone preparations.

The history of endocrinology offers numerous examples of the fact that animal experiments were usually the key to the discovery of hormonal processes and thus provided the basic facts which clearly would not otherwise have been obtained. Some particularly important examples may be highlighted: Berthold (1848), with hens, and Knauer (1896) with rabbits, prevented the consequences of castration by testicular and ovarian implantation. Both researchers rightly suspected that this was effected via the blood stream - as we now know, by hormones. This is how models of hormone action in humans were developed, thanks to further very detailed animal experiments. Mering and Minkowski (1898) showed that the surgical removal of the pancreas in dogs produced diabetes. In 1921, Banting and Best discovered that an extract from dog pancreas reduced the high blood sugar level of pancreatectomised dogs as well as of human diabetics, providing the first effective treatment for diabetes. Of the 20 or more steroids which Reichstein had extracted from adrenal gland extracts during the 1930s, one was shown in animal experiments to be biologically effective; it was the most valuable of the corticosteroids. The search for thyroid gland hormones was only possible by testing metabolic effects on dogs with and without their thyroid glands. Pituitary gland hormone research is almost unthinkable without experiments involving their surgical removal (hypophysectomy). Furthermore, Haberlandt, in the years after 1919, found in his experiments on rats and guinea-pigs that the corpus luteum of pregnant animals is able to temporarily interrupt the fertility of related species, and in doing so, laid down the basis of modern oral contraceptives. We owe these and many other significant discoveries in endocrinology to the correct interpretation of animal data and to intelligently devised animal experiments. Such data has given great support to the hypothesis that, in certain organs, substances are produced which exercise an important biological function remote from their place of origin.

Qualitative hormone determination
Once the existence of hormones had become evident, the need arose to isolate and identify these substances. For this purpose too the animal experiment was very important, since at that time, the existence of hormones could only be demonstrated by determining their biological action. The identification of hormones, the structure of which may be either relatively simple or very complex, was for a long time impossible by physico-chemical methods, and had to be achieved by observing their pharmacological actions *in vivo* (bioassays). In addition, it was mainly through the development of reliable hormone bioassay systems that the isolation of these natural substances, in the amounts needed for structural analysis, was accomplished. Hormones are generally so highly active that they are present in the body in extremely low concentration. As the following examples demonstrate, their isolation often involves concentration exceeding one million fold:

- more than 100 l of urine from pregnant animals were needed to obtain 1 gram of choriongonadotropin.
- 250,000 sheep hypothalami were used to, produce 1 mg of thyrotropin-releasing factor.
- About 500 kg of adrenal cortex, taken from 20,000 animals were needed for the isolation and identification of less than 1 gram of pure corticosteroid.
- The ovaries of 50,000 pigs were used for the first isolation of 20 mg of pure corpus luteum hormone (progesterone).

Thus, purifying a hormone is frequently very difficult, and the necessary purification steps needed to isolate the substance could only be verified on the basis of the biological activity of the hormone. For this reason, suitable biological detection methods had to be developed for the isolation of the hormones.

The isolation of hormones using biological test procedures has contributed to the clarification of their structure, and has also rendered possible their synthetic manufacture and the synthesis of chemically altered compounds with therapeutically desirable effects. Indeed the latter is of great medical significance since the use of chemically modified hormone-like substances which have no undesirable side effects may be therapeutically advantageous. Thus for example, scientists developed orally-active oestrogens and progestagens, adrenal cortex hormones with a specific effect on inflammatory processes or on mineral balance, the relatively non-androgenic anabolics and modifications of a hypophyseal posterior pituitary hormone, which possess either mainly vascular or diuretic effects, as desired. Such developments were and are still only possible with animal experiments, as many hormonal effects can only be detected in the bioassay. Among the countless examples are the essential differentiation of antiphlogistic (inflammation inhibiting) and mineralocorticoid (affecting the mineral balance) effect of adrenal cortex hormones and their artificial modifications, the blood sugar-lowering effect of insulin and ovulation inhibition by steroidal contraceptives.

Quantitative determination of hormones
1. Animal experiments
Biological test procedures can be used not only for the qualitative determination of hormone actions but also for quantitative hormone determination. For this purpose, the results are expressed in appropriate bioassay units. Because of the countless, almost uncontrollable factors which influence bioassay units (differences in animals, weight, feeding, or ambient temperature etc.), the need soon arose to develop a reliable reference system. This was achieved by using so-called biological standards. In principle, the biological action of the test material is compared with the action of an international (or national) standard for the substance in question. The results of the procedure can then be expressed in international units (IU). The standard is a highly purified product of the biologically active substance (e.g. the purest insulin contains 24 IU per mg, and is produced by the prescribed method under the auspices of the World Health Organization (WHO). The standard preparation is then stored by the WHO and may be obtained from the biological standards committee of the WHO). The international unit is then defined as the activity of a given weight of the standard (e.g. 1 IU of insulin corresponds to the action of 41.67 micrograms of the 4th international standard preparation). The action of the test material is compared with that of the standard under standard bioassay conditions (e.g. 1 IU of

insulin lowers the blood sugar from 6.66 to 2.22 mmol/l in a 2 kg rabbit which has fasted for 24 hours). In this way, many sources of error are eliminated and reproducible results may be achieved. Incidentally, the bioassay for the quantitative determination of insulin is painless to the rabbit, and corresponds exactly to the clinical process of establishing the insulin dose in diabetics by blood sugar determinations.

Hormone bioassays have since mostly disappeared from clinical endocrinological diagnosis. The physico-chemical methods described in the next section allow, almost without exception, the measurement of hormone concentrations in blood or urine. For example, laboratories which previously used thousands of toads or frogs every year for pregnancy testing, now measure the pregnancy-induced increase in urinary gonadotropins by an immunological method.

It is quite different in endocrinological research. New hormones, particularly neurohormones, were discovered because of their biological action. To isolate the hormone this action must be measurable. However, even after its isolation, structural analysis and synthesis, a specific assay method must be available. Hormone synthesis often facilitates the subsequent production of chemically similar derivatives with the object of obtaining compounds which possess a greater or more selective action than the naturally occurring hormone. This has happened with most steroid hormones and some peptide hormones. It is clear that comparisons of efficacy can only be made on the basis of biological activity. Examples of available methods include:

(a) Determination of organ weight after the animal's death (e.g. seminal vesicle weight of rats after injection of a male sex hormone).
(b) Blood level determination (e.g. determination of the adrenal cortex hormone in rats after injection of the hypophyseal regulatory hormone ACTH).
(c) Isolated organs (e.g. the posterior pituitary hormone oxytocin in the isolated rat uterus).

Thus, there is no substitute for using research animals in endocrinology during the research and certain development phases. Evan at later stages, animal experiments cannot be completely relinquished because a number of biologically important, even vital hormones, could only be developed for therapeutic use, in a form with known and constant activity, on the basis of biological methods. Examples are insulin, gonadotropins and the growth hormone.

Thus, animal studies in endocrinology research have not only been very significant in the development of the specialty, but remain extremely important. Their elimination would impede progress in many areas and hinder the production of very valuable drugs.

2. Physicochemical methods

Animal-based bioassays are often more sensitive and specific than physical and chemical methods, but are generally used much less than these techniques. They are usually considered more expensive, inaccurate and prone to error. Efforts have therefore been made to replace bioassays with physicochemical methods; efforts which have already been successful in many areas. Hormone synthesis and analysis by physico-chemical methods sometimes permits the omission of the biological tests previously required, and the substitution of bioassay units by units of weight. For example, adrenocorticotrophin (ACTH), a pituitary hormone, is a protein of 39 amino acids, and is hardly detectable chemically in drug production. Dosages are expressed in international activity units. In contrast, synthetic ACTH, a molecule with only 24 amino acids, offers, together with other advantages, easier chemical analysis and dosages in units of weight.

On the other hand, the developments of chemical, radiochemical, gas chromatographic and radioenzymatic, but primarily those of radio-immunoassays (RIA) and of competitive protein binding assays, have superceded many biological tests (see also p. 105). The measurement of most known hormones in biological fluids is now mainly performed with highly sensitive radioimmunological methods. For some largely scientific problems however, biological procedures are still used because levels of a protein measured by biological and immunological methods are not always identical. Thus, the antibodies may be directed against a part of the protein which is not essential for biological activity, or the antigenic site may be missing in biologically active fragments. In addition, where the discrepancy is great, the results of biological assays usually hold good. Even with RIA, animals cannot be completely relinquished, since they are needed for the production of antibodies, although not for the test itself. Because of their importance, the principle of immunological assay methods will be briefly described. Firstly, animals (rabbits, sheep, goats and guinea-pigs) are immunised with a highly purified preparation of the hormone to be determined. Thus, antibodies are generated in response to the appropriate antigen; in this case the hormone. Subsequently, dilutions of the antibody-rich serum are mixed with the material in which the presence of the target hormone is to be measured. The resultant antigen-antibody reaction follows the law of mass action, and is thus dependent on the concentrations of the hormone and the specific antibodies. It can be manifested in various ways. One method uses not only antibodies against a highly purified hormone, but also a radioactive preparation of this hormone; the radioisotope ^{125}I is used to label proteins, and ^{3}H or ^{14}C to label steroids. If the radioactive hormone comes into contact with the antiserum, a radioactive antigen-antibody complex is formed. If this reaction occurs in the presence of the test material containing the unknown amount of hormone, then the latter will also react with the antibodies. The non-labelled ('cold') hormone will thus compete with the labelled ('hot') hormone for binding sites on the antibody and displace a part of the radioactive hormone from the complex. In this way, some of the labelled hormone remains in solution, and cannot be bound to the antigen because the binding sites are occupied by the non-radioactive hormone. Thus, the amount of non-bound radioactivity is proportional to the concentration of hormone in the test fluid. The relationship between bound and free radioactivity can then be established by physico-chemical methods, and the hormone concentration of the test sample measured by reference to an appropriate standard curve.

The labelling of the hormone can also be achieved by binding to an enzyme, by which a typical colorimetric reaction for the enzyme is inhibited. Such methods include EIA (enzyme immunoassay), EMIT (enzyme multiplied immunological test) and ELISA (enzyme-linked immunosorbent assay). Further hormone labelling methods which do not involve radioactive isotopes utilise fluorescence or luminescence.

In summary, immunological determinations for hormones have replaced expensive and inadequate animal research in wide areas of endocrinology including pregnancy testing, and have made biological methods superfluous for many purposes.

References

Burn, J.D., Finney, D.J. and Goodwin, L.G. Biological Standardization. Oxford University Press, Oxford 1950.
Gaddum's Pharmcology, 8. Aufl., von Burgen, A.S.V. and Mitchell, J.F. Oxford University Press, Oxford 1978.
Labhart, A. Clinical Endocrinology, 2nd. ed., Springer, Berlin 1986.
Schuurs, A. and van Weemen, B. Enzym-immunologische bestimmungsverfahren. Diag. Intensivther. 4, 2 (1979) 17-21.
Szpilfogel, S.A. Adrenocorticol steroids and their synthetic analogues. In: Parnham, M.H. Discoveries in Pharmacology, vol. II. Haemodynamics, Hormones and Inflammation. 1984 (p. 253-284).
Tausk, M. A brief endocrine history of the German-speaking people. In: Kracht, J., von zur Mühlen, J. and Scriba, P.C. Endocrinology Guide, Federal Republic of Germany. Dtsch. Ges. f. Endokrinologie. Brühlsche Universitätsdruckerei, Geissen 1976 (S. 1-34).
Tausk, M., Thijssen, J.H.H. and van Wimersma Greidanus T.J.B.: Pharmakologie der Hormone, 4. Aufl. Thieme, Stuttgart 1986.

The 'lonely' heart
Cardiovascular research
H. Juan and G. Raberger

Cardiovascular research is one of the most important sections of experimental and clinical pharmacology (Fig. 9). Heart and blood vessel diseases (e.g. myocardial infarction, atherosclerosis, hypertension etc.) are important causes of death. There is no doubt that the considerable recent advances are, or will be, of great therapeutic significance (e.g. new understanding of the origin of atherosclerosis, mechanism of action of calcium antagonists etc.). Nevertheless, there are considerable gaps in our knowledge of the relationships of cardiovascular function to other physiological systems. For example, completely new impulses are emerging in neurophysiology (e.g. the discovery of specific neuropeptides, which together with the classic transmitters, may play a role in regulating vascular function) and fatty acid research (e.g. the possible anti-thrombotic role of a fatty acid in some fish oils). This means that further animal experiments are always required.

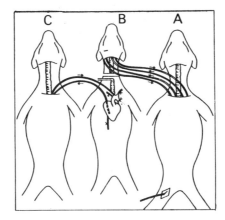

Figure 9. *Three dogs for one research project:* Dog A serves only to supply blood from its heart to the head of dog B. The heart of dog C perfuses dog B's heart which is isolated from the circulatory system. Only through these complicated experiments was Heymans (1927) able to demonstrate the reflexes through which cardiac function is adapted to performance. For this work, he was awarded the 1938 Nobel prize.

Anyone who today has his cardiovascular and lung function painlessly tested in a clinical laboratory - with and without exercise stress - should be grateful to these experimental dogs. Without these early studies, modern clinical diagnosis could not have evolved. To some extent, a wide range of so-called 'alternative methods' has been available in cardiovascular research for decades. These techniques are being successfully used, and are constantly being supplemented by new methods for specific problems. However it must be stressed that complex processes, such as effects on blood pressure, cannot be examined by these 'alternative' *in vitro* methods. Although in the final analysis, the effect of drugs on the cardiovascular system must always be tested on the whole individual (animal *and* man), almost all the preliminary work can now be performed on *in vitro* systems.

To some extent, *in vitro* methods are used in drug development, particularly in basic research. In addition, routine investigations, such as the testing of anti-anginal drugs on the isolated heart are quite possible. Of the utmost importance in experimental research is the understanding of the functions, the study of these functions in pathological processes, and the search for the mechanisms of action of pharmaceutical agents.

What methodological possibilities are currently available? As the following examples show, they are by no means limited to the *'lonely'*, and thus the *'isolated'* heart, but are considerably more comprehensive:

1. Isolated portions of the guinea pig heart such as vestibule and papillary muscles.
2. Isolated hearts of different animal species. The classical models are frog heart (Straub) and the isolated mammalian heart (Langendorff). Newer variants have been developed from these.
3. Isolated blood vessels of different species including man.
4. Isolated perfused organs such as heart (see above), kidneys, lungs, skin etc.
5. Isolated single cells or heart or blood vessel tissue (endothelium, smooth muscle) or of the heart.
6. Cells (e.g. erythrocytes, platelets) of the haematological system.
7. Subcellular particles or tissue homogenates.

<u>Point 1.</u> Vestibule preparations are partly used to measure the action of compounds which influence the contractile strength of the heart. Agents of this kind play a major role in the therapy of cardiac insufficiency. Influences on heart beat can also be readily observed in these muscle preparations.

<u>Point 2.</u> Isolated whole hearts are somewhat more complex to use. The oldest preparation of this kind is the frog heart (Straub) which is now relatively seldom used. The toxic effects of potassium ions and cardiac glycosides are particularly well demonstrated.

The <u>mammalian</u> heart provides much more extensive data however. In the original experiment (by Langendorff) the heart was removed after the death of a laboratory animal. The coronary vessels were then perfused with a nutrient medium through an aorta cannula.

In this way the heart remains functional, and continues to beat rhythmically for hours. The heart's frequency and contractile strength, as well as the flow through the coronary vessel can be measured. This design is not only useful for routine drug testing, but has been introduced into basic research programmes. A much more recent example should illustrate this. The isolated guinea pig heart is extremely well adapted to promote the understanding of the mechanism of anaphylactic reactions. Completely new stimuli for in-depth study of these processes have resulted from the discovery of leucotrienes. These are endogenous compounds which are produced by allergic processes and, in asthma for example, cause a constriction of the airways, and participate in the anaphylactic heart reaction. The latter is characterized by severe constriction of the coronary vessels, decrease in contractile strength and the appearance of arrhythmias. Under certain conditions (e.g. sensitization) the reaction can be triggered in isolated guinea pig hearts. Histamine, leucotrienes and other compounds, which are rapidly produced in large quantities, are responsible for these symptoms. The effects of the production, release, and action of the mediators, as well as the course of the reaction are readily investigated with this experimental system.

This technique can be further improved. Thus for example, the nutrient solution can be conducted through the left vestibule, enabling both heart performance as well as myocardial oxygen requirement to be measured. Furthermore, the composition of the nutrient solution can be improved; whole blood or blood components (erythrocytes, platelets, proteins etc.) can also be used.

Point 3. Isolated vessels such as aorta (rats, rabbits), mesenteric vessels (rabbits), coronary vessels (cattle, pigs) and various human blood vessels are considered suitable for investigating physiological mechanisms. Contractions and relaxations, mechanisms of vasoactive agents, biochemical processes, drug interactions and similar processes can be examined. At present, attention is being focussed on the importance of the endothelium (the innermost cell lining of vessels) for the action of endogenous compounds, the mechanisms of atherosclerosis ('vessel calcification') and the influence of certain fatty acids on vessel function.

Point 4. Apart from the isolated heart, other organs (e.g. kidneys, spleen, liver, hind legs, ears etc) may be used, which are removed from a freshly killed animal and perfused with nutrient medium through an artery. Variations of flow or perfusion pressure, as well as production and release of endogenous compounds can be investigated.

Point 5. Isolated vessel cells (endothelial or smooth muscle cells), as well as heart cells, are ideally suited for biochemical investigations of the reactions of individual cells.

Electrophysiological experiments may also be performed, as is exhaustively described on p. 86.

Point 6. The same applies here as for point 5. At present, blood platelet reactions and their significance for different vascular diseases are being studied particularly intensively.

Point 7. Homogenates or subcellular particles from other organs may also be used for biochemical experiments.

The heart of the intact organism
It is beyond doubt, that with the methods so far described, the available data on the function of the *whole* organism, as well as the action of drugs on the whole organism are inadequate. It is therefore quite erroneous to accept that *interactions* of the cardiovascular system with other organs, and their regulation by the nervous system can be accurately assessed by studies on isolated organs. The heart and blood vessels are subject, as are all the other organs in the body, to continuous control. It is therefore guaranteed that cardiac function will adapt to the requirements of other organs. The heart can be viewed as a pump in a closed branched system of tubes. Both the pump (the heart) and the tubes (the vessels) can receive via several *feedback mechanisms*, biological feedback from the organs which rely on the blood supply. This feedback occurs on the one hand via the nervous system, and on the other via the blood stream. Heart rate, i.e. the number of beats per minute, as well as cardiac performance, i.e. the amount of blood pumped per beat, is increased or decreased by the nervous system. Furthermore, the amount of blood flowing into different areas is regulated by neural control of the vessels. Feedback from these organs via the blood can take place in the heart as well as the vessels. Numerous endogenous compounds are involved, which are produced in different tissues or organs and which influence heart rate, cardiac function and the diameter of the vessels.

If these associations have been understood, it will be obvious that every assertion about the cardiovascular system is valid for man only if the investigations have been carried out on the whole body, i.e. on the intact experimental animal. This applies equally to normal physiological regulation as well as to the function of the diseased heart. By analogy, drug testing is also only meaningful if it is performed in a similar disease situation, but with a fully-maintained regulatory system. It can be deduced from the above considerations that not only the integrity, but also the initial state of the cardiovascular system is of paramount importance for this kind of study. Compounds which affect the response of the heart, or indeed of the vessels, can thus interfere with both normal physiological associations and the clinically relevant testing of vasoactive drugs. The resting function of the heart and the vessels are equally influenced by anaesthesia, and the function of the feedback mechanism is also markedly reduced.

In modern cardiovascular research, efforts must therefore be made both to analyze pathophysiological associations, and to test drugs, under clinically relevant pathophysiological conditions. *If possible*, this should be performed on the conscious animal. It goes without saying that the animal must be completely free of fear or anxiety, and this can be achieved by training. This is absolutely essential for a meaningful outcome to the experiment. As far as the stress affecting the conscious experimental animal is concerned, it is a general rule that *everything which is performed on humans without anaesthesia is also acceptable in animals.* This would certainly apply to all non-invasive measurements, i.e. cardiovascular tests, which utilise measuring probes outside the body. Invasive methods also apply however, e.g. catheter examinations, which are necessary for measuring heart function and pressure ratios in vessels. The effects of active drugs on blood pressure, heart rate, cardiac output etc. can be studied with these techniques, which correspond to clinical methods used for circulatory diagnoses, or the supervision of a patient in intensive care.

Theoretically, there is much to be said in favour of so-called computer simulations. In principle, the computer is programmed with the basic functions of the body, and for example, can calculate the effect of a drop in blood pressure due to a drug acting on the blood flow of different organs. Although this is possible in principle, it has no clinically relevant value, since the computer only possesses that knowledge which the user has programmed into it. Furthermore, the computer cannot distinguish why, for example, the blood pressure falls. However, for teaching purposes this approach is quite important. The complicated interaction of different feedback systems with the cardiovascular system can never be taken over by a computer; it can only be simulated with the help of data previously acquired in experimental animals or humans (see p. 106).

Small enemies - great dangers
Chemotherapeutic research in animals
H. Obenaus and J.G. Meingassner

In recent decades, considerable advances in chemotherapy have been achieved. Numerous chemotherapeutic agents are available to modern doctors, and the lack of these drugs would be unthinkable. Great success has been attained in the battle against the pathogens of infectious diseases, those 'small enemies' of man and animals, i.e. parasites, fungi, bacteria and viruses, which often lead to severe, contagious and fatal diseases (Fig. 10).

The successful control of infectious diseases was achieved partly through *immunization* as prophylaxis against such diseases as smallpox or poliomyelitis, and partly through the development of synthetic or antibiotic compounds which, because of their bactericidal or bacteriostatic effects, are suitable for *chemotherapy* or *chemoprophylaxis*. The latter has been particularly useful in the control of infections caused by bacteria, fungi or parasites, and finds its expression in a wide range of highly effective preparations. With parasitic infections such as malaria, eliminating the carriers (mosquitoes) offers an additional effective measure to avoid the risk of infection.

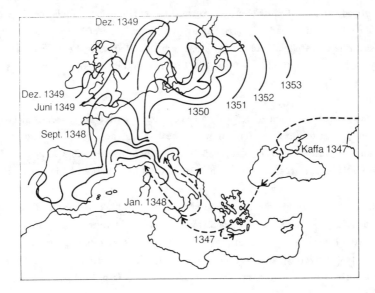

Figure 10. *The plague travels around Europe.* In Sicily in January 1348, and in northern Norway in December 1349, the plague pathogen (*Pasteurella pestis*) lived in rat fleas. Rats with infected fleas are the prerequisite for the plague in man, and must be present in his immediate environment. Humans infected with this disease contaminate others directly. At a time when the ship and the horse were the quickest forms of transport, the plague epidemic spread through the whole of Europe in three years, leaving depopulated areas in its wake.

In an age when 50 million human beings sit closely packed in aeroplanes, we are barely conscious of the value of hygiene, infection prophylaxis, inoculation and chemotherapy. Even less do we appreciate the fact that this safety for mankind is due to experimental animal research in these areas.

What is chemotherapy?

Prerequisites for the targeted development of chemotherapeutics, based on the epoch-making work of Paul Ehrlich in the second half of the 19th century, are sufficient knowledge about the pathogens involved, and the development of procedures which allow these pathogens to be grown in the laboratory, or if this is not possible, in *infected animals*. In this way, compounds can be tested for the desired effect. The aim of treatment is either to inhibit the multiplication and growth of a pathogen, or to kill it. In general, this type of testing is performed *in vitro*, and involves exposing a culture of the organism to a specific concentration of the test compound for a fixed period. From these experiments, the degree of activity, or by testing against different pathogens, the spectrum of action of a compound, can be established. Although a compound may be effective *in vitro*, this is no guarantee that it is also active *in vivo* in infected humans or animals. For *in vivo* activity, the injected compound must come into direct contact with the pathogens and be available in infected tissues, in adequate concentration for a sufficient length of time, to kill the pathogens or to impede their growth. Thus, some compounds which are highly active *in vitro*, fail in experimental animals *in vivo*, because of insufficient resorption, rapid excretion or inactivation due to transformation or degradation, or because the compound binds to host tissue, preventing sufficient accumulation of the drug in the infected area. A further uncertainty factor is that growth requirements for pathogens in body fluids and tissues are different from those in synthetic nutrient media. Thus, it would be irresponsible to test experimental drugs on seriously ill patients without first testing them in animal experiments, even if these agents are shown to be potentially chemotherapeutic by *in vitro* test systems which are still being devised, and by computer models.

This type of testing may be performed in laboratory animals where the compounds can be studied under similar conditions to those anticipated in man. The most reliable prediction for efficacy in man comes from experiments where the laboratory animals are infected with a pathogen which is infectious to man, where the experimental infection leads to a disease which broadly corresponds to the human condition, and where the kind of treatment which is foreseen in man also leads to cure in animals. In addition, the physiological data from laboratory animals, concerning resorption, distribution, metabolism and excretion of compounds, so far as this is possible by analogy, should largely correspond to those of humans. Infection models with pathogens which are equally infectious for both man and animal, usually present no difficulties.

However, there are numerous pathogens which are certainly infectious in man, but only lead to illness in certain infected individuals, often in unknown circumstances. In these cases it is not easy to simulate the disease conditions in laboratory animals. Often the choice of a specific species of experimental animal is the key to a successful study. There are examples however, where instead of human pathogenic organisms, one has to use pathogens which cause a disease in experimental animals corresponding to the human infection. Naturally, the quality of an animal experiment also depends on the degree of its standardization and the reproducibility of the results.

In this way the development of numerous valuable chemotherapeutics has been made possible, such as the antibacterial sulphonomides, penicillins and cephalosporins, the antifungal agents amphotericin B and ketoconazole, the antiparasitics resorcin, ivermectin or praziquantel and the antiviral drug virazol. In spite of these major successes it is erroneous to suppose that the development of chemotherapeutics is a closed subject. Indeed, as much scientific effort as ever should be directed at the development of new chemotherapeutic agents. Thus, the fact that the efficacy of specific agents may be inadequate, infectious pathogens may become resistant to drugs after a time, or diseases (e.g. AIDS) may not be controllable with available agents or by immunization, indicate the need for new preparations.

Testing chemotherapeutics in animals involves all the precautionary investigations (see pp. 41 and 56-59) which must be undertaken for all groups of drugs. For example, one must consider that the chemotherapeutics used on humans or animals may not only destroy the disease pathogen, but may also affect the functions of the 'host organism'. Many compounds which are very active *in vitro* prove to be too toxic in initial animal tests and must be discarded. However, how can this be recognised without chronic toxicity testing on the whole animal before proceeding to clinical trials?

An impressive example of the necessity to develop new drugs against different pathogens is provided by the prophylaxis and treatment of malaria.

An example. Malaria therapy and prophylaxis
Long before it was realised that malaria is not caused by toxic vapours from swamps (mal aria = bad air), but by microorganisms which enter the human blood stream via a mosquito bite, the therapeutic and prophylactic efficacy of quinine was known. The Spanish conquerors of Central America discovered the curative powers of Peruvian bark from the local natives. The colonization of tropical and subtropical regions around the world showed the incredibly wide spread of this epidemic, from which thousands of millions of people suffer. White colonies could only protect themselves against it by continually taking quinine. Peruvian bark trees were secretly exported to South East Asia, and plantations were established to satisfy the increasing demand for quinine - a piece of medical history which reads like a detective story. The Frenchman Laveran discovered the malaria plasmodium, the Britons Ross and Manson demonstrated their transmission by mosquitoes, and the Italian Grassi introduced the first hygiene measures for their control.

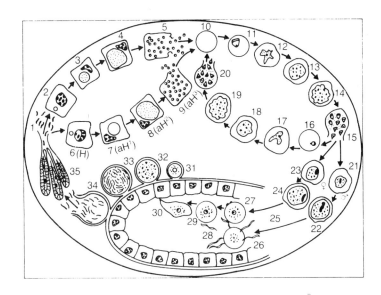

Figure 11. Development of *Plasmodium vivax* (from Peters and Richards; Antimalarial Drugs 1, *Handbuch der experimentellen Pharmakologie*, Vol.68/1. Springer, Berlin 1984). 1-24 Developmental stage in man, 25-35 developmental stage in the anopheles mosquito, 1: Sporozoites from the salivary gland of the mosquito, 2-5 and 6-9: pre-erythrocyte phases in liver cells, 2- 4: formation of exoerythrocyte schizonts, 5: mature tissue schizonts with emerging merozoites, 6-9: formation of hypnozoites, 10: attack on erythrocytes by merozoites and activated hypnozoites, 11-14: formation of erythrocyte schizonts, 15: mature blood schizont with emerging merozoites, 16-20: repeated merozoite formation in blood schizont, renewed attack on erythrocytes by merozoites, 21-24: formation of micro-and macrogametocytes, 25-35: formation of gametes, zygotes, oocysts and infectious sporozoites.

In 1891, Ehrlich and Guttmann reported the successful treatment of two malaria patients with methyl blue, and for the first time demonstrated the possibility of a causal therapy of infectious diseases by chemical compounds. However, in spite of the progress and interim successes since these early origins of chemotherapy, the control of malaria is still a pressing problem. The efficacy of available drugs is becoming lost because *Plasmodium falciparum* strains, the pathogens of the worst form of malaria, have become resistant to them. In addition, there are no effective forms of immunization.

The importance of infection models and their contribution to the development of chemotherapeutics are forcibly demonstrated in this example (Fig. 11). The transmission of the plasmodium to man takes place via infected female *anopheles* mosquitoes while sucking blood (Fig. 11). Sporozoites get into human blood together with the mosquito's saliva, and subsequently enter the liver where they multiply in a pre-erythrocytic phase into tissue schizonts to produce merozoites. These merozoites then penetrate the red blood cells, again forming schizonts (blood schizonts) and dividing into merozoites. Thus, a cycle of repeated fevers occurs, since the merozoites released from the rupturing erythrocytes repeatedly re-parasitise the red blood cells (erythrocytic phase). Finally individual merozoites differentiate into gametocytes, which are taken up by blood-sucking mosquitoes and then become sporozoites, ending the transmission cycle.

It is clear that implementation of drug therapy in this complex cycle of developments is not easy. Four different human malaria pathogens exist, which vary in their specific stages of development, and react differently to drugs. Thus, to interfere with the developmental cycle in man, effective drugs are necessary, which either kill the parasites before they multiply in the blood cells, or which prevent their increase in the erythrocytes and in their maturation to gametocytes (blood schizonticides). Furthermore, in the case of *Plasmodium vivax*, the second most important malarial pathogen, hidden developmental stages may be found in the liver for months, or even a year (so-called hypnozoites). In spite of the most intensive efforts it has not so far been possible to reproduce the developmental cycle which occurs in the human body from human pathogenic *Plasmodia*. Indeed, for many years, even the long term culture of asexual intraerythrocyte phases was not possible.

Thus, so much more sensational were the methods of W.Trager, J.B. Jensen and J.D. Haynes, published in 1976, for the continuous cultivation of *Plasmodium falciparum* in red blood cells (blood schizonts). For the first time, this method enabled researchers to study infected blood samples *in vitro*, to conduct large scale *in vitro* tests of chemical compounds against blood schizonts of the pathogens most dangerous to man, and to obtain material for immunization trials. In spite of this progress, recently boosted by the successful cultivation of further developmental stages of *Plasmodium falciparum* and *Plasmodium vivax* in tissue cultures, the development of new drugs has, as previously, relied on animal experiments.

Initially, pigeons, hens and budgerigars infected by avian malaria were used to test new compounds. In 1948, it was discovered that free-living rodents could also contract malaria. Subsequently, mice, rats or hamsters, which had been experimentally infected by rodent malaria, were used in preference to birds because the malaria pathogens of birds and mammals are very different, and rodents are biologically closer to man than are birds. This explains why results of avian studies were not always applicable to man. The major advantage of using rodents was that the experimentally-induced disease closely resembles the course of human *Plasmodium falciparum* infection, and this enabled the complete developmental cycle, i.e. the transfer of malaria by mosquitoes to the mouse and vice versa to be successfully established in the laboratory. This model made it possible to design experimental conditions to assess the ability of compounds to combat the different developmental stages of the malaria pathogen. These test systems led to the development of several anti-malarial drugs, which was wrongly interpreted as a solution to the problems of prophylaxis and therapy of malaria, and led to a belief that malaria could be eradicated.

However, the 1960s brought disillusionment. Not only had mosquitoes developed resistance to the most varied insecticides, but strains of *Plasmodium falciparum* appeared which were not susceptible to resorcin, the most important and frequently used agent. Resorcin resistance was noted in all malarial regions and represented a new challenge to medical research. Efforts to find new effective agents or possibilities for protective immunization also led to new test models. It was discovered that Douroucouli monkeys and Squirrel monkeys could be infected with *Plasmodium falciparum* and *Plasmodium vivax* enabling the direct *in vivo* testing of drugs on these important human pathogens, as well as the start of immunization trials. In recent years, the compounds Fansidar and Mefloquin have been developed and brought into use. However, even after a short time, malaria pathogens which had become resistant to Fansidar were being isolated from patients. The appearance of Mefloquin-resistant strains has since also had to be contended with, because animal experiments have shown that parasites which are resistant to this drug, simultaneously develop a reduced sensitivity to other agents. Thus, the battle against malaria seems to require the continuing development of new preparations. It is like a constant alternating battle as the focus changes from man producing a new agent, to the parasite developing a resistance. In this respect, animal research is of inestimable value since it permits 'in the field' development. One knows from experience that the parasites develop a resistance by repeated under-dosing of infected animals. Thus it is possible, even before a drug is widely used in man, to develop a follow-up preparation which might be effective against strains which have developed a resistance to the previous compound.

It would be desirable, by developing an effective vaccine, to avoid the constant struggle of searching for new drugs. However, despite intensive efforts, and the introduction of the most modern technology such as cell fusion for the production of monoclonal antibodies, or genetic techniques, no practical success can yet be envisaged.

As the malaria example shows, animal experiments are an unconditional requirement for chemotherapeutic advances. Without animal experiments the development of new chemotherapeutic agents is impossible. However, it is no longer animal experimentation alone, but the increasing number of new procedures for breeding pathogens *in vitro* or in tissue culture, which, even more than animal experiments, facilitates the selection of new compounds. Thus, animal experiments are restricted to far fewer, but more positive models.

Today, anyone going on holiday in a malarial region should be grateful, not just to researchers, but also to their experimental animals. Perhaps he may even realise that people could not live there at all without appropriate chemoprophylaxis.

References

Bruce-Chwatt, L.J. Malaria, the growing medical and health problem. Drugs exptl. clin Res. 11 (1985) 899-909.

Garnham, P.P.C. The present state of malaria research: a historical survey. Experimentia 40 (1984) 1305-1310.

Peters, W. and Richards, W.H. Antimalarial drugs. vol. I. II. In: Hanbuch der experimentellen pharmakologie. Band 68/I. II. Springer, Berlin 1984.

Wagner, W.H. Malariabekämpfung: Chemotherapie und immunoprophalaxe. 1., 2. und 3. Mitteilung. Arzneim-Forsch./Drug Res. 36 (1986) 1-9, 163-175, 409-415.

Safety for the healthy - treatment for the sick
Cancer research
R. Schulte-Hermann and W. Paukovits

In the countries of the industrialised world, cancer has become one of the most frequent causes of death. Statistically, every fifth death is a result of cancer. At present, the battle against cancer must be conducted in two areas:

1. Protection against carcinogenic factors.
2. Treatment and cure of cancer patients.

Effective protection against carcinogenic factors requires that we are able to identify these with certainty. The difficulty in recognising these factors is not appreciated by most lay people. Indeed, broad sections of the public already believe that the 'poisoning' of our food and environment with numerous synthetic chemicals is responsible for many cases of cancer. Similarly, the detection of pesticides, antibiotics, fertilizers and hormones etc. in food, soil, water and air, constantly cause alarming reports in the media. Are these substances, which we largely have to thank for our present hunger- and infection-free existence, responsible for the terrifyingly high incidence of cancer? Is cancer the price we have to pay for our wellbeing? This chapter will show how science seeks to answer these questions, and the extent to which animal experiments are necessary, or can be replaced by other methods.

There is now no doubt that genetic factors may play a significant role in certain types of cancer, but for the majority of human cancer patients, *exogenous* factors are actually important causes of cancer ('cancerogenic' or 'carcinogenic'). Amongst these, specific chemical compounds, as well as radiation (radioactive radiation and the ultraviolet element of sunlight) undoubtedly cause cancer in man and animals. Certain viruses are also carcinogenic - in research animals, and (probably) in man. When one considers that to date, about 7 million different chemical compounds have been synthesised, of which about 60,000 are actually used and can therefore act on man and the environment, then the extent of this toxicological problem is clear. Nature supplies a similarly large and so far barely known 'reservoir' of chemicals.

For the evaluation of the carcinogenic (and otherwise toxic) potential of a compound, it is not important whether it is chemical or natural in origin. Natural substances still hold the record for toxicity. For example, the poison Aflatoxin, a product of mould, frequently contaminates foodstuffs in countries with inadequate food hygiene and supervision. This substance can cause cancer in experimental animals and (probably) in man, and is far more active than the so-called 'supertoxin' Dioxin (2,3,7,8-TCDD). Cycasin too, a component of flour made from the fruits of certain palms (Cycas), is a carcinogen which may contribute to the increase of liver cancer in some tropical countries. In addition, extracts of birthwort (*Aristolochia*) from our native flora, were used as medicines until a few years ago when a severe effect causing stomach cancer was observed in rats.

What procedures are currently available for the identification of carcinogenic factors? The oldest method uses epidemiological and statistical studies. With these, attempts are made to show connections between living- and food-consumption patterns, as well as occupational factors on the one side, and the incidence of specific kinds of cancer on the other. Thus, the cancer-producing action of soot and tar in chimney sweeps has been known for over 200 years. Appropriate hygiene measures in this group of workers resulted in 'tar cancer' disappearing in the subsequent decades.

So far, epidemiological studies have identified with certainty 30 chemical compounds and mixed compounds as human carcinogens (International Agency for Research on Cancer, Lyon 1985). However, the potential of epidemiology is limited on several grounds:

1. Cancer usually first appears years or decades after contact with the carcinogen, the type of which can often no longer be ascertained after so long a time.

2. Cancers in man almost never have a **single** cause. Even tobacco smoking, the most important and clearly identifiable human carcinogen, only leads to cancer in a small percentage of smokers. In general, the action of a carcinogen does not necessarily lead to the onset of cancer, it only heightens the statistical risk. In this sense, even tobacco smoke is 'only' a 'risk factor', which, primarily in conjunction with other such factors

only leads to the observed incidence of cancer as a cumulative effect. Less significant risk factors are found only with difficulty.

3. Epidemiology can only indicate a statistical relationship (correlation) between the incidence of disease and specific exposure. As this can also be due to chance, a causal connection can only be detected by appropriate experimentation.

The second oldest way of identifying carcinogenic factors is the animal experiment. All human carcinogens, which have been adequately investigated to date, may also be shown to induce cancer in experimental animals. Using animal experiments, the carcinogenic effects of several hundred further substances have been recognised, in many cases before they were used. There is no doubt that we must thank animal experiments (and the sacrificed experimental animals) for the fact that some catastrophes have **not** occurred. One thinks for example, of aromatic amines such as dimethylaminoazobenzene, and of nitrosamines etc. Unfortunately, the results of animal experiments have not always been heeded. Thus, the carcinogenic effect of diethylstilboestrol (DES, a synthetic hormone with oestrogenic action) was known, but it was still used for 20 years to prevent miscarriage in pregnancy before it was noticed that the children of treated women developed malformations and cancer of the genital region.

Despite the incontestable value of animal experiments, efforts have long been made to develop practicable, *in vitro* ways of identifying carcinogenic compounds. This is supported on both ethical and practical grounds, i.e. the long duration (3-4 years) and high cost of animal experimentation (in 1986, the cost of a carcinogenicity trial on rats or mice was about 20 million Austrian schillings or 1.4 million DM per substance). The difficulty of developing suitable *in vitro* tests lies in the fact that cancer appears at the end of a years-long, multi-phase and complex process, which previously could not be reproduced *in vitro*. Basic research to elucidate the individual molecular and biological dynamics leading to the transformation of healthy cells into cancer cells in the living organism, offers the only way out of this dilemma. Not until these dynamics are known can they be studied and tested *in vitro*, to find out if and how they are influenced by specific chemical agents, irradiation or viruses.

Perhaps the most significant advance in fundamental cancer research in recent years is the recognition that many chemical carcinogens, as well as cancer-inducing radiation and some viruses, damage the genotype, and thus can give rise to mutations (also known as a 'genotoxic' effect). Amongst the subsequently developed *in vitro* procedures for the detection of genotoxic effects is the 'Ames test', named after its inventor (see p. 83), in which mutations are identified in bacteria. In other tests, yeast cells or warm-blooded cells are used, as well as flies (*Drosophilia*). Other methods of testing are based on chemical changes in the genetic material (DNA) itself. As a corollary, it may be noted that even the Ames test and other *in vitro* methods cannot manage without animal material, namely liver enzyme preparations, since most carcinogens can only damage the genotype after enzyme-catalyzed 'activation'.

Unfortunately, it was soon apparent that these genotoxic tests **cannot** identify carcinogenic substances with certainty. Recently the 'International Commission for Protection against Environmental Mutagens and Carcinogens' reviewed the accuracy of genotoxicity tests as indicators of carcinogenicity in the 280 chemicals for which sufficient information was available. There was a 'hit rate' of 80%. Most human carcinogens are also mutagenic and thus detectable in these tests, although there are some significant exceptions, e.g. DES and asbestos. Although genotoxicity tests are undoubtedly an important step towards replacing experiments on living animals (and on man), they are not reliable predictors of carcinogenicity on their own.

A further breakthrough seemed to have been achieved when cultured, non-malignant cells were successfully transformed into cancer cells by carcinogens *in vitro*. Unfortunately however, 'cell transformation tests' are so far technically difficult, lengthy and rather unreliable; the 'hit rate' is still lower than that of genotoxicity tests. A great deal of work, time and money still has to be invested in this approach. As before, a tumour-inducing substance can only be defined from the fact that it actually creates a tumour, and this evidence must be deduced from animal experiments.

This unsatisfactory situation led to more attention being paid to the multifactorial and complex nature of the development of cancer. Animal studies and observations in man show that certain substances promote the development of 'pre-cancer' stages. Since pre-cancerous stages are probably relatively numerous, exposure to such tumour-promoting substances can also lead to cancer. It has therefore been realised that chemical substances must also be tested for tumour-promoting effects. However, the question is: how? At present, the mode of action of causative agents is poorly understood. To some extent, reliable *in vitro* tests still yield nothing in spite of intensive efforts. To date, the only available tool has been animal experimentation, although definite progress is being made. The activity of tumour-promoting agents is beginning to be discernable by the growth of *pre*-cancerous stages, and before the onset of malignant tumours. This certainly demands considerable work. However, because pre-malignant stages appear much more frequently than tumours, and thus increase the statistical accuracy of the results, a drastic reduction in the need for experimental animals will be possible (to about 20% of the requirements of a 'normal' carcinogenicity study). A **decrease** in the number of experimental animals should be a sensible interim objective towards full-scale abolition of animal experimentation.

Can 'safety for the healthy' from the cancer-inducing effects of chemical substances be achieved today with animal experiments? Essentially we must accept that there is no **absolute** safety and never will be, with or without animal experiments. However, modern toxicology has developed test strategies with which the greatest possible safety can be attained. The heart of these strategies is a hierarchically structured programme, in which *in vitro* methods and animal experiments are seen not as 'alternatives' but as being mutually complementary. Every new compound is first subjected to the genotoxicity tests described above. Substances which prove positive in these tests are always relegated to the chemists' archives. Even if the compound has passed genotoxicity and other toxicity tests without giving concern, and even if its metabolism and excretion from the organism are known, it is still tested in an animal experiment for carcinogenesis and further delayed effects. Only a small percentage of the compounds synthesized reach this latter testing period, before widespread use in humans or in the environment, only decreasingly small amounts of the test compound are used. However, at this point in the test programme the animal experiment is quite indispensable. Substances which have passed all previous tests may still be detected here as being unexpectedly cancer-inducing. By the logical combination of all available means, including animal experiments, a 'hit rate' of almost 100% can be anticipated. Abandoning animal experiments would mean that new carcinogens would come on to the market every year. Furthermore, cancer research would be robbed of one of its most important tools, and thus the development of new or improved *in vitro* tests would be halted.

How does the role of 'chemicals' now stand in the above-mentioned increase in malignant diseases? First it should be understood that this increase is largely due to our longer life expectancy (most types of cancer occur more frequently in higher age groups), for which we have to thank the protection against other causes of death, such as infection, starvation, polluted foodstuffs etc. Chemical agents, especially components of tobacco smoke, have contributed to an increase in the incidence of cancer. Substances used in the production and preservation of foodstuffs, and in pest control etc, as well as other chemicals in recent commercial use, are well studied toxicologically. Nevertheless, a 'residual risk' may still be attached to them and can never quite be eliminated. The large number of 'old compounds' which were developed before the introduction of modern toxicological tests represents a problem, as does the presently inestimable abundance of natural substances. Following earlier experiments, some unpleasant surprises may still be facing us here. The adequate testing of these numerous compounds is a challenge for future toxicology which can only be dealt by using **all** scientifically useful methods.

Despite considerable research effort, we still do not know the causes of most human cancers and therefore cannot protect ourselves from them. Particular importance is therefore attached to efforts to combat the clinical manifestations of tumours as effectively as possible. There are currently two ways to attain this goal:

a) The optimization of existing forms of therapy.
b) The development of new methods of treatment and drugs.

For various reasons, not every one of the numerous forms of therapy in existence represents a viable way to manage every patient's illness. One cancer is not necessarily like another even if it occurs in the same organ; it is only a slight exaggeration to say that almost every patient has his own individual cancer. On the other hand, no known form of therapy is without considerable side effects, which can demonstrate sharply individual differences.

Tumour chemotherapy is aware of numerous cell growth-inhibiting agents (cytostatics), a considerable number of which have found their way into clinical practice. While, from years of international experience, the choice of a particular drug or drug combination for specific tumours can be made with reasonable accuracy, because of the individual differences mentioned above, there are frequent problems in the choice of optimal therapy for an individual patient.

Here, valuable help can be provided by techniques developed in recent years. Using tissue culture methods, it is now possible to grow the tumour cells of many patients *in vitro* (tumour stem cell assay) and to determine, after relatively short term culture, to which drug the afflicted patient will predictably react positively and to which he will not. Although the method has not fulfilled the hopes originally placed in it - unfortunately an agent which is very effective *in vitro* is often not so in clinical use - it does usually give reliable results for non-efficacious drugs. By culturing the individual tumour cells of a patient it has thus become possible, by avoiding ineffectual treatment strategies, to gain valuable time, and to reduce discomfort for the patient.

Only in relatively few cases will it be necessary, in order to determine suitable chemotherapeutic agents, to grow the tumour cells in 'nude mice'. These animals have specific genetic defects (e.g. hairlessness), and it is particularly significant that they have lost the ability to respond to foreign cells with any defensive reaction. This immunological deficiency makes it possible to transplant human tumour tissue into the animal and to investigate the efficiency of therapeutic measures. The potential for using this procedure is considerably restricted by the fact that these animals are difficult to rear and maintain, and because of their immune deficiency, they have to be kept in special germ-free conditions.

With the development and testing of new drugs, test tube studies in tumour therapy are gaining increasing importance. The number of experimental animals in drug research has, according to the latest statement from the German Association of Pharmaceutical Industries, dropped from almost 4.2 million in 1977 to 2.4 million in 1984. A portion of this reduction is attributable to the fact that it is now possible, by breeding primary tumour cells *in vitro*, to forego animal experiments for the initial tests on most new substances. Only a small proportion of the compounds tested mainly for their action against cancer cells gets beyond this first stage into a research phase in which, to achieve their aim, additional animal experimentation cannot be avoided. Since animal and human tumours can react very differently to cytostatics, the use of nude mice carrying human tumours, is a particularly important tool in the development of new cytostatic drugs.

This also applies to every clinical investigation which is carried out for 'genotherapy' of cancer. Tumour-producing viruses have been known for decades, but it is only a relatively short time since it was discovered that in the Rous sarcoma virus for example, the suppression of a single gene (of the oncogene *src*) results in the loss of tumour-inducing activity. The intensive research resulting from this observation has within the last decade led to a completely new understanding of the mechanism of tumour occurrence. In the course of this work it was established, totally unexpectedly, that genes are also present in non-infected cells, which are closely related to *src* and other viral oncogenes.

This finding led to the discovery of so-called proto-oncogenes and cellular oncogenes, in which often a very slight change (mutation), for example the exchange of individual nucleotide bases, is sufficient to activate the tumour-inducing potential. At present, the most widely accepted concept of 'activated proto-oncogenes' is that the proto-oncogenes in the healthy organism perform important functions. However, through changes brought about by the introduction of a tumour virus (viral oncogene) or even through other virus-independent agents (cellular oncogenes), the proto-oncogenes are so altered and activated that uncontrolled growth of the affected cell takes place. Activation of proto-oncogenes by the genotoxic action of chemical carcinogens seems to be an important step in cancer production mediated by these substances.

In the meantime, more than 30 oncogenes have been discovered, and *in vitro* methods play a significant role in explaining their mechanism of action. For example, cells which can only normally grow *in vitro* after becoming attached to a firm base, grow even without attachment after 'transformation' by oncogenes (anchorage independent growth). This is a signal for the formation of so-called transformed phenotypes *in vitro* and acts as an important indicator for the transformation of a normal cell to a tumour cell. Nevertheless, neoplastic transformation can only be proved by implanting these cells in a suitable host organism and subsequently detecting actual tumour growth (combined *in vitro*/ *in vivo* assay).

Most of the oncogene-encoded proteins discovered so far play a role in the regulation of cell division processes and gene activities, be it as growth regulating factors, or as specific cellular receptor proteins for such factors. Thus, the protein encoded into the genetic code of the sis-oncogene is structurally altered by an important growth factor (Platelet Derived Growth Factor, PDGF), while the proteins belonging to the *erb*B and *fms* oncogenes show similarities to those cellular receptors which recognise and react with certain other growth factors such as Epidermal Growth Factor (EGF) and Colony Stimulating factor (CSF). Many oncogenes play a role in the modification (phosphorylation) of proteins, which perform important functions in the cellular regulatory system. Evidence of the cellular function of some oncogenes and oncogene-proteins is still lacking. Together, the findings to date quite clearly show that cellular oncogenes occupy key positions in the regulatory system of cells, which can cause serious disturbances upon structural and functional changes. The importance of oncogene-coded cellular receptor proteins in such control mechanisms again points to the role played by the interaction of individual cells in the whole organism in respect to their functional integration and their disturbance in cancer.

The refining of oncogene detection methods should ultimately make it possible to identify changes in this gene in human DNA, and as a result, to routinely test neoplastic cells from patients. This should open up new ways of elucidating the mechanisms by which tumours originate, and of selecting effective forms of therapy for them.

Justified hope for new ways to treat tumours lie in the possible exploitation of the molecular differences between malignant and normal tissue. Thus for example, specific antibodies could be directed against the protein of an activated oncogene (assuming that it differs from 'normal protein'). 'Genotherapy' in the true sense, in which direct changes in the genome of the neoplastically transformed cells can be produced, is theoretically possible in the light of more recent oncogene research. However, this can only be realised if the 'team work' of the genes within a cell, as well as of cells within the whole organism in controlling growth, is adequately understood.

Molecular biology is adequately served with the first of these problems by *in vitro* methods, although the integration of all regulatory phenomena in the control system of the whole organism cannot be analyzed without using the 'whole organism'. Thus, the research situation at present, and for the immediate future, seems to be characterised by the fact that the use and significance of *in vitro* methods in all areas of cancer research are increasing sharply. Nevertheless, it is impossible to draw conclusions about the situation in the whole organism without investigations on the intact organism, and therefore without animal experimentation.

The fight against cancer
Manoeuvres with biological and chemical weapons?
H.H. Sedlacek

The power of man over nature, and the extent to which man exploits it, as well as the threat to existence, have today ensured that life as something worth protecting has become a theme of central interest. More significant, although not the only motivation, may be the knowledge that one's own existence depends on the viability of the environment.

Very much older than the present discussion about environmental protection is the debate about protecting animals. This is partly associated with the species-dependent mystical theory of the incarnation of divine strength and power in ancient Egypt, partly on the species of animal and partly also on the individual animal, associated in Hammurabi's book with promoting the protection of useful species - of animals which man particularly loves. An affinity with the animal which is not species-dependent is clearly combined with the teaching of reincarnation as it has survived in Hinduism, Buddhism and Jainism. Consideration for all living things, compassion for all creatures and the prohibition of killing them is related to the theory of reincarnation in Buddhism and Jainism. This teaching of reincarnation was the basis for the school of Pythagoras in Europe before the time of Socrates, which forbade animal sacrifice, the eating of flesh and 'injury to animals' as handed down by Xenocrates.

According to Roman law by contrast, the animal was an object with no rights and was treated accordingly. This attitude to animals predominated until the 19th century in the treatment of animals in central Europe. The Roman legal attitude to animals was complemented by the man-centred (anthropocentric) theology of Thomas Aquinas, according to which "the animal, since it has no immortal soul, cannot be a person, and since it is not a person, there is no obligation to accord it justice or love". Over the centuries, the ethical considerations of distinguished philosophers were only oriented to this anthropocentric way of thinking.

The Roman legal attitude to the animal remained undisputed for a long time until the arrival of Francis of Assisi. His relationship to animals was indeed founded on a supernatural point of view, but the ethic of brotherhood he propounded demanded care and compassion for all creatures capable of suffering. Schopenhauer also recommended unconditional compassion for all living things, since this would be the best surety for moral good conduct.

Animal experiments had previously been performed on living animals as part of displays at fairs and public exhibitions, to gratify onlookers wanting a gruesome public spectacle, for example by showing beating hearts and quivering organs. In 1871, the German penal code first virtually banned these public demonstrations of 'animal experiments' on aesthetic grounds, but scientific experiments on living animals remained permissible.

Today's Christian attitude to the animal is rooted chiefly in the Jewish religion. The foundation of the Jewish as well as the Christian attitude is the biblical requirement "subdue the earth and rule over the fish of the sea, over the birds of the sky and over all animals which walk on the land"; further, "so the Lord God took man and set him in the Garden of Eden, so that he would cultivate and guard it", and the invitation to Noah; "all living creatures which move shall be sustenance for you" and further the stipulation of love and charity to animals, as exemplified in the picture of Noah's ark and in the sparing of the city of Nineveh.

According to today's Christian ethical interpretation, animal life is distinct from plant life. Animal life possesses its own particular value, this being in a living community with mankind, but it is only expressly man, and not animal who makes an adequate partner for man and is in the likeness of God.

Considering the subordination of the animal world on the one hand, and the command to protect it on the other, Albert Schweitzer only countenanced animal experiments if, after examining their necessity in each individual case for the maintenance of life and the health of man, they are unconditionally necessary.

Böckle (1984) stated the fundamental ethic that "life is to be protected in accordance with the ethical values of respect and love in the observance of the principle of responsibility". Here however he distinguished between human life and animal life, not in the sense of "worthwhile life against worthless life, but human life against other lives possessing their own worth". The participation of man in the mandate of creation means, according to Böckle (1984), the obligation to promote life, and thus too the authorisation of man to sacrifice animal life where necessary in order to save, promote, protect and preserve human life. This service of animal to human life is however, subject to two conditions: the minimisation of pain for the animal kingdom and a sense of proportion for the evaluation of the animal's wellbeing.

The animal protection law
From this markedly Christian moral basis, Pastor Albert Knapp founded the first animal protection society in Germany in 1837. One hundred years later (1933-1938), this was followed by the animal protection law. This was at a time when numerous German citizens would have regarded it as an act of mercy to have the protection of this animal protection law themselves, while Hitler aspired to an unconditional ban on animal experimentation.

The trauma of the Third Reich (Nazi rule in Germany) taught us that looking at the torture of animals as a moral wrong (because it renders human brutal and insensitive to human suffering) may be quite correct, but without a corollary; the most brutal guards in the concentration camps could be considerate towards their dogs, and the politics of murder entailed the goal of completely abolishing animal experiments.

The German animal protection law of 1933 aimed to protect the life and wellbeing of the animal, and under the general aegis of a ban on painful or harmful animal experiments, allowed only unavoidable animal experiments, and these only under the supervision of specific solicitous and caring conditions. Furthermore, it limited the number of experimental animals to that needed for reaching the stated goal. This law, and the supplementary regulations introduced in subsequent years, underwent a testing time not only during the period of the Third Reich, but also in the ensuing years of democritisation and economic prosperity. For reasons of increasing practical knowledge, the newly formulated animal protection law (1972) was supplemented by the introduction of the obligation to notify the authorities. All animals are protected from unauthorised use in scientific studies, whereby experiments on vertebrates must be officially approved if the same result can not be achieved by research on cold-blooded animals.

This law is regarded by noted scientists and animal protectionists as important to control essential animal experiments. To quote B. Grzimek: "Our whole field of medicine - the successful control of so many diseases and infections in man and animals - has been made possible by animal experiments. It is pointless to close our eyes to it. On the other hand, we can justifiably demand that animal experiments are only undertaken with anaesthesia, never unnecessarily, and only with responsible scientific supervision".

However, there were and still are opposing views, e.g. those which mainly regard today's animal experiments as being justified from both scientific and ethical standpoints, and who see it as the goal of any civilised and normally sensitive person to have mastery over all animals. A corresponding change in the animal protection law is envisaged as a central socio-political problem. At the root of the initiatives to amend the animal protection law lie two basically differing viewpoints:

The principle of equality and the balancing of rights
The demand for the 'principle of equality in the true communion of nature' came from Meyer-Abich (1984). According to this, animals are not analogous to man, but do have equal rights. In particular, the principle of equality between man and animal in their capacity for suffering, and the observance of the interests of specific animal species, is proposed as the basis of a new natural legitimate partnership.

While the principle of equality in capacity for suffering is already allowed for in our present animal protection law, inasmuch that no painful operation may be performed on a vertebrate without anaesthesia (unless an anaesthetic is not required for a comparable human operation), there are doubts about the possibility of considering animal-specific interests. Indeed, many indicators lead one to suppose that animals experience the world in ways which are different to those of man, and that even from one species of animal to another, their spheres of experience are quite differently ordered. Thus, for man it has long been impossible to describe these species-specific worlds in terms of his own perceptions, let alone objectively place the protection of these specific worlds within the legislative framework.

Apart from other evidence, the long-recognised differences in the neurophysiological and the perceptive-physiological organisations indicate, that, unlike man, other mammals are less aroused by optical impressions, but are considerably more so - although to some extent this varies from species to species - by the stimuli of smell, taste and touch. Inadequate awareness of this by the animal's owner may often cause pain and suffering, especially in domestic animals and pets. However, to publish detailed legal provisions within the framework of animal protection, would not be helpful because of our fragmentary knowledge of how the animal perceives its world. It is obvious that we are able to objectively determine animals' sensitivities to pain. Pain of this kind inflicted on animals is considered in the animal protection law. Whether we are capable of perceiving all the sensitivities of the animal to pain, for example through which sense channel it is stimulated, is questionable. In this respect, the pain sensitivities in man and the higher primates seem to be too different from say, rodents on the one hand, and carnivores on the other.

Using the principle of equality in the legal defined community of nature, the one-sided references to man enshrined in the existing legal policy between man and animal should be broken down and dissolved (Meyer-Abich 1984). This is characterized by the various forms of subjugation of animals, which are represented in designations such as hunting animals, zoo animals, pests, parasites and disease-carriers, sacrificial animals, useful animals, slaughter animals, domestic animals and pets, and experimental animals. As a first result of the demand for equality, particularly in relation to the capacity for suffering, a ban on all animal experiments is proposed, which are not also allowed in humans. In contrast, killing animals for human food is expressly permitted. Compared to depriving an animal of its freedom, or injuring it physically and/or psychologically, the killing of a living creature is surely the highest measure of disenfranchisement. On the other hand, the slaughter of an animal is a deliberate act by man against the animal, just as is its use for basic research, and both these deliberate acts can be regarded as equal in man's use of the animal. The claim for the principle of equality between animal and man is thus in contradiction to the simultaneous sanction of the slaughter of animals practised by man. Any subsequent implementation of the principle of equality between animal and man would be to concede for animals and for man at least the right to life, to physical integrity and to freedom to the extent that other living creatures are not encroached upon. This ruling would have to apply to all animals, irrespective of their usefulness to man; this includes pests and parasites.

In this connection, the rearing and maintenance of domestic animals and pets, of experimental and domestic animals, would be questioned and would have to be re-evaluated. The hunting of wild animals and the slaughter of domestic animals would be improper. It is obvious that any subsequent implementation of the principle of equality between man and animal would be a burden to man.

If one restricts the discussion about the principle of equality to animal experimentation for tumour research, then, testing for carcinogens using genotoxic (mutagenic) substances on experimental animals, and inducing or implanting tumours in specially-bred experimental animals to evaluate new methods for understanding and treating tumours, would not be allowed either because it is not permissible on humans. Thus, man would have no substitute on which to test diagnostic methods and drugs before their clinical application, and thereby recognise toxic side effects before they could be harmful to humans. Had the principle of equality existed previously, we would today have no anti-tumour drugs at our disposal, since the actions of all the known anti-cancer agents were discovered through animal research. In addition, the number of carcinogen-induced human cancers would probably be considerably higher than it is today, since the evidence of carcinogenicity of many substances obtained through animal experiments would not exist.

In contrast to the principle of equality, the balancing of rights represents the ethical basis for animal research. This balance is favoured by, and based on, a series of philosophers and sociologists such as R. Spaemann, O. Höffe and G.M. Teutsch, as follows:

The dominion of man over animals may be accepted, but the right of man to carry out animal experiments is confronted by the right of the animals to life and physical integrity. Since man, by his mental ability and his self-control, is superior to animals, he has a moral responsibility to consider the animal's capacity for pain and anxiety.

Without doubt, the right of the animal to life and physical integrity is an ethically grounded value. On the other hand, animal experimention to improve the diagnostic and therapeutic potentials of human and animal diseases is also ethically-based. In individual cases, a decision for animal research can only be taken by weighing one of these morally-grounded prerogatives against the other. An animal experiment planned in this way is therefore ethically defensible and thus permissible. This particularly applies to physical injury, the infliction of pain, and deprivation of freedom. Inflicting pain on an animal cannot be justified for any other purpose other than the avoidance of comparable pain or the saving of life.

From the balancing of these rights it follows that animal experiments are basically not ethically sanctioned, unless the rightful claim of man is seen as being superior to that of the animal, and the animal's claim is considered insofar as the experiment is restricted to the most essential and the infliction of pain to the absolute minimum. However, measurement of this minimum is in dispute. Thus, there are opinions which recommend avoiding all animal experiments which cause the animal severe pain. In this connection, severe pain is seen as that condition which man would consider to be intolerable without alleviating measures.

What would have been the consequences of observing this recommendation in cancer research? It is known that in most cases, advanced human cancer leads to unbearable pain, which makes medical intervention necessary. Is it therefore not permissible to induce in animals such as mice and rats, corresponding tumour diseases by implanting tumour cells or by administering toxic substances, in order to test the effects of new cancer treatments on these animals? If so, we shall today have to adjust our research towards new, urgently needed drugs for tumour diseases. If this recommendation had been valid in the past, none of the present recognisably effective anti-tumour drugs obtained by animal research would have been discovered. Nevertheless, today approximately 5-8% of all tumour diseases can be cured by these drugs, and prolongation of life through their use is anticipated in 40% of cases. We must now renounce these medical successes. This for example shows the problems of restricting the balancing of rights to arbitrarily appointed guidelines.

In the opinions of a number of philosophers, moralists and lawyers (R. Spaemann, F. Böckle and E. von Loeper), animal experiments which purport to test coffee, tea, cigarettes or other such items for their harmlessness should be abandoned. This recommendation is not to be attributed entirely to the scientifically accepted balancing mentioned above, between the rights of man and the rights of animals, but must rather be understood as an indicator to the legislators to shape the legal directives for safety testing using animals so that it is possible to strike a balance. Within this balance, testing for the carcinogenicity of a new substance found in coffee, tea and cigarettes for example, which had been shown to be genotoxic *in vitro*, or in which there is the danger of metabolism to a toxic substance, may be fundamentally admissible.

It is clear that this balancing in the planning of animal experiments must be undertaken by specialists; experts whose ultimate aim from the medical standpoint is the diagnosis, therapy or prevention of human disease; experts who, from the veterinary point of view, can assess the type and degree of anticipated physical damage, the pain inflicted and other stresses on the experimental animal, and who can appeal, within the terms of the animal protection law, on behalf of the legal interests of the animal.

It will serve neither the rights of men nor those of animals if the balance in respect of any one project falls under the influence of political parties, ideologies, philosophies or economic political interests. Proposals for changing the existing ratification procedures for animal experiments are in this sense to be evaluated for the good of sick humans and for the best possible preservation of the rights of experiment animals.

Criteria for further changes in the animal protection law should be concerned with the interests of both humans and animals. Idealistic demands, for example the ban on animal experiments are not concerned with the logical call for animal experiments for the benefit of man. Both objectives could be legislated for, but are seemingly incompatible. Nevertheless, a way must be found which reconciles current scientific demands with moral and ethical norms. This is assuming that some moral ideals are not taken to be generally obligatory norms.

Animal experiments in fundamental research
Tumour research also constitutes a major portion of modern basic research, in which various types of specialty, such as pharmacology, toxicology, virology, molecular biology, immunology and cell physiology, together with their specialist affiliations to human and animal medicine participate. In the past, these interrelationships have provided information which contributes greatly to the presently tentatively successful measures to prevent, diagnose and treat human cancers. To these belongs the relatively complete understanding of the manifold causes of the incidence of cancer:

- its causation and/or furtherance by carcinogenic substances (carcinogenesis), viruses, cancer-promoting substances (promoters) and other collaborative and mutually reinforcing factors.
- the dependence of tumour incidence on heredity (genetic factors).
- the molecular and biological analysis of these genetic factors to attain clearly defined genetic information.
- the evidence of the changing (transformation) of a normal cell into a malignant cell capable of producing a tumour by transmission of several genetic substances whose actions supplement each others'; awareness of the toxification and detoxification of carcinogens in the living organism with regard to their direct or indirect interaction with the genetic material.

From the study of the body's defences (immunology) we have derived:

- knowledge about intracellular (cellular) and fluid (humoral) antibody-mediated immune reactions.
- information about the cells participating in this immune reaction, such as phagocytes (macrophages), immune cells (B- and T-lymphocytes) and active substances (e.g. interleukins and interferons).
- evidence of cell-destroying (cytotoxic) immunological defence reactions against tumour cells.
- an understanding of the ways in which tumour cells can evade immune defences.

This work in tumour immunology clarified the incredibly varied and rapid ways in which tumour cells can - depending on the surrounding milieu - alter their outer coating, i.e. their cell membrane pattern of substances (antigens) which can be recognised by the immune defences as foreign. Implantation studies have shown that, irrespective of whether it originates from one cell (monoclonal) or from many cells (polyclonal), a tumour node comprises a large number of tumour cells, which are capable of behaving very differently. These differences include the development of surface characteristics (cell membrane antigens), growth rate, the capacity to create secondary growth (metastases), and their susceptibility (sensitivity) and ability to withstand (resistance) the various tumour cell-destroying drugs (cytotoxics) or the individual immunological defence mechanisms.

This diversity (heterogeneity) in tumour cell characteristics within a malignant tumour, and the changeability (variability) in the behaviour patterns of tumour cells, shows the considerable problems in treating tumours. The objective of this therapy is to destroy those tumour cells which remain after surgical removal of visible tumour from the patient, and which subsequently lead to renewed tumour growth (recurrence) or to secondary tumours (metastasis).

Much of this knowledge about tumour biology, which could be described here only very briefly, was obtained through animal experiments. Many of the results of virological, biochemical, biomolecular, immunological or pharmacological (i.e. non-animal) tests have gone into planning this kind of experimentation, or these methods have been a necessary supplement to experiments on tumour-bearing animals. A prerequisite for such tumour experiments on animals is the genetic production of homogeneous experimental animals. These are inbred lines in which for example, tumours can be reproducibly induced by toxins or radiation, or transplanted without rejection reactions, or which are immunologically incompetent and thus permit the growth of human tumour cells.

Despite our knowledge of tumour biology from animal experiments, we are still very far from solving the problem of neoplastic disease. However, without this knowledge, and the advances already achieved in the prevention and treatment of tumours, cancer would be much more of a danger to man than it is today. Further urgently needed successes in tumour therapy, be it painstakingly obtained in small advances, or with a major 'breakthrough', are only possible by continuing with basic research. Their findings serve as the basis for the development of new drugs, even if these preparations (as has happened often enough) are discovered not only by systematic testing but also by 'accident'.

The prevention (prophylaxis) of cancer
At present, prophylaxis of neoplastic disease is mainly restricted to avoiding the uptake of cancer-inducing agents. To date, carcinogens have been mainly discovered in long-term experiments (carcinogenicity studies) using animals specially suited for this work, and to whom the suspect compound was administered. Indeed, the results of these investigations, mainly performed on mice and rats, could not always be directly extrapolated to humans because of animal-dependent toxicity or detoxification mechanisms, although this circumstance does not necessarily diminish the value of animal experiments.

Newly discovered or manufactured compounds can in fact be classified as 'suspect' by *in vitro* experiments, e.g. using tests for genetic impairment (mutagenicity tests) and malignant transformation (transformation tests) in bacteria and cell cultures. However, the carcinogenicity of these substances still has to be determined in appropriate animal experiments. The protection of both man and animals against potentially cancer-inducing natural products or chemicals can only be effectively achieved by an understanding of these agents, and thus only with carcinogenicity studies on suitable experimental animals. The ethical foundation of these studies is provided by the goal to be attained.

Without animal experiments, we would only be able to recognise the carcinogenicity of suspect substances retrospectively on the basis of statistical results from human diseases. In humans, this method would not yield sufficiently rapid, plentiful or reliable data about the danger of natural substances or chemicals because of the delay, often spanning decades, between uptake of the carcinogen and the onset of the neoplastic disease. If one restricted oneself to the clinical findings, despite knowing that the carcinogenicity of suspect substances might be detected more quickly by rather more predictive animal experiments, one would be guilty of neglecting the care of one's fellow men. Furthermore, it would be tantamount to choosing man as the test subject for testing cancer-inducing compounds.

In addition to avoiding the ingestion of carcinogens, other preventive measures, which might protect man from the effects of carcinogens should be considered. These may include for example, vaccination against those tumours which are virus-induced, or even the administration of agents which accelerate or enhance the deactivation of carcinogens. Vaccines, for example to prevent certain tumours in chickens, or the hepatitis-virus vaccine against human liver cell tumours, have already demonstrated efficacy or are being tested. However, for other means of protection the groundwork remains to be performed, and performed too with experimental animals.

The treatment of tumours
Surgical excision and radiation therapy of tumours are well established methods of treating neoplastic diseases. However, the success rate of these methods in combatting cancer in the secondary growth stage (metastasis) are very limited. As a result, metastatic spread is the most frequent cause of death in tumour patients.

Treatment with anti-tumour drugs (cytostatics) was developed for the management of leukaemias, micrometastases and advanced tumours which cannot any longer be treated by surgery or radiotherapy. In addition, numerous other forms of tumour therapy are currently being sought for and tested. In this way, work in the field of tumour immunotherapy has made the greatest progress.

Anti-tumour drugs (cytostatics)
There is now an imposing array of clinically effective cytostatics, all of which have been tried and tested on animal models. Thus, their anti-tumour efficacy on selected implanted tumours (leukaemias, melanoma and cancer of the lung, colon and breast) in the mouse (less so in the rat) has been compared with their general toxic effects (toxicity). On the strength of these selective methods, drugs which show a sufficiently wide anti-tumour spectrum in patients with assorted tumours have been studied. Consequently, the cytostatics undergo extensive long-term clinical studies, either individually or in combination, and often in comparison with other possibilities for tumour therapy. To date, the results of clinical studies are encouraging on the one hand, but disappointing on the other.

It is encouraging that chemotherapy prolongs life in about 40% of cancer patients, and results in a cure in some 5-8% (e.g. in over 50% of childhood leukaemias, in women with cervical cancer and in men with testicular cancers). Nevertheless, it remains disappointing that with most tumours, particularly those of the alimentary canal and lung, but also with other tumours, the growth of micrometastases or clinically detectable tumour nodes is unaffected, or only transiently affected by cytostatics. In these areas, a prolongation of life, if it occurs at all, is only very slight and is influenced by the toxic side-effects of cytostatic treatment.

The success of current cytostatic tumour management is incontestably based on animal experiments. Without these experiments, which on a worldwide basis have been most extensively and consistently performed by the National Cancer Research Center (National Cancer Institute) in the USA, with more than 400,000 test compounds (even now about 10,000 new substances are tested every year), we would not now have the potential to cure or alleviate some human cancers by cytostatic therapy. A critical comparison of animal experiments with clinical findings however, shows that cytostatics chosen for their effectiveness against certain kinds of experimental tumour, sometimes turned out to be completely ineffective against the same kinds of tumour in man. Thus, there is still no cytostatic drug which would for example, be definitely effective for colonic neoplasms, although numerous agents produce regression of experimental colonic cancer in mice.

In view of the great number of drugs already tested, there is relatively little chance of the future discovery, in the course of all the routine tests on cytostatics for experimental tumours, of any which are clinically effective against such human tumours for which we do not have any therapy at present. An alternative or supplementary means is suggested by the method used for testing the cytostatic efficacy of a drug on human tumours. This technique was originally developed to distinguish, prior to administration, effective cytostatics from ineffective ones for the management of an individual cancer patient by using his tumour as test material. Inhibition of the capacity of tumour cells to divide (taken from the tumour and placed into cell culture) is used to measure cytostatic capacity in therapeutically relevant and clinically tolerated concentrations. In several studies, it has been shown that the predictive ability of this system is greater than all other methods, i.e. it is possible to forecast effectiveness in more than 50% of cases, and ineffectiveness in more than 80%.

However, the use of this approach for routine studies of new cytostatics is complicated by the limited availability of human tumour material and the widely differing susceptibility to cytostatics between tumours of the same variety, but in different patients. A solution to this problem is provided by the use of continuous human tumour cell cultures, or preferably by the introduction of continuously growing human tumours in nude mice. These tumours grow after transplantation of suitable surgically-removed material in nude mice, since these animals have genetic defects in their immune defences and cannot reject foreign tissue. Both sources offer standardizable, routinely available cell material in essentially unlimited quantities. However with human tumours grown in nude mice, via serial inoculation from animal to animal, the diversity (heterogeneity) in the character of the tumour cells, originally obtained surgically, is far more easily retained than it is in continuous cell cultures. Heterogeneity becomes reduced in the course of culture transfer, mostly to cell strains with only one (monoclonal) or fewer (oligoclonal) parent cell lines. The heterogeneity in the composition of the test cells also influences the results of cytotoxicity testing: false-positive results occur far more often in continuously cultured tumour cells than in cells freshly obtained from human tumours in nude mice. On this basis, *in vitro* cytotoxicity testing has been established for a greater number of human tumours, grown in nude mice, as a new approach to the discovery of new cytotoxics. Substances which have shown growth-inhibiting activity on human tumours in the test tube, are tested at concentrations which are both attainable and tolerable in the organism, for anti-tumour effects on the animal. i.e. on the corresponding human tumours grown in the nude mouse, in order to exclude results which are restricted to the test tube and which cannot be followed up in living organisms. Testing on human tumours in the nude mouse demonstrates above-average predictability for clinical studies. If necessary, the anti-tumoural efficacy in standardised experimentally-transplanted mouse tumours is ascertained in parallel with test tube and animal testing on human tumours. Drugs with activity against both animal and human tumours, as well as agents whose effects are limited to human tumours, are clinically tested on patients with appropriate neoplasms. The results so far indicate that there are drugs which are ineffective on standard experimental mouse tumours (including leukaemia), but have a distinct anti-tumour activity on human tumours (e.g. leukaemias and gastrointestinal tumours).

Accordingly, it is possible that the search for cytostatics in experimental mouse tumours has so far been inadequate, since drugs which have been ineffective here could have been quite efficacious on human tumours. Because of this fear, even the American Cancer Research Center has supplemented its experimental animal test system for finding new cytostatics, by including *in vitro* cytotoxicity tests on human tumour cell cultures, and tests for anti-tumour efficacy on human tumours in the nude mouse.

Altogether, the number of studies on experimental animal tumours may possibly be reduced by the inclusion of *in vitro* experiments, but this reduction is at least partly outweighed by the use of nude mice with genetically defective immunity for the production of human tumour material, and for the subsequent testing of cytostatics on human tumours in the mouse. New procedures in this area of tumour research, including those for the creation of alternative methods, result in further animal research. A voluntary curtailment or ban on animal experiments would drastically hinder or prevent the construction of new predictive test systems for finding urgently needed tumour therapeutics.

Tumour immune therapeutics
Experiments to influence the growth of tumours by activating the immune system are more than 100 years old. Latour (1851) and Busch (1866) had already attempted to influence tumour growth with bacterial infections. From these beginnings stemmed experiments, particularly in the 1920s and 30s, and subsequently intensified in the 1960s and 70s, to find a new approach to the treatment of tumours via the immune system. These efforts were supported by the explosive increase in knowledge about cellular and humoral components

of the immune system, mechanisms for recognition and information processing of foreign antigens, regulation and deregulation of immune response and the possibilities for antigen-specific (against only one foreign substance) and antigen-unspecific (against many foreign substances) defence of tumour cells. All this information came primarily from experimental animals. The use of experimental animals also resulted in the discovery and testing for potential anti-tumour properties of various mediators of the immune response (e.g. interferons, interleukins and tumour necrosis factor).

Clinical trials were planned on the basis of these animal experiments. Thanks to genetic engineering methods it was possible to produce a series of human immune system mediators in adequate amounts for clinical testing. So far, the results of these clinical trials, e.g. with α-interferon, show clear evidence of clinical efficacy for some tumours, such as hairy cell leukaemia or some of the kidney tumours, but no definite therapeutic effects could be detected for the majority of tumours. Because of these results, the initial euphoria of possessing new broad-spectrum weapons for treating human tumours, in the form of these mediators, has given way to disillusionment. Current expectations for the present clinical trials of mediators (γ-interferon, interleukins and tumour necrosis factors) have now been reduced in the belief that the therapy of previously untreatable tumours can only be achieved by a major breakthrough.

This disillusionment should not however lead us to think that the clinical successes in tumour therapy so far achieved with mediators, are less than those attained with many cytostatic agents. Furthermore, and parallelling the experiences with cytostatics, it remains to be seen whether combinations of mediators with each other and with other anti-tumour agents can improve therapeutic success.

Far more disappointing than the results with mediators have been the numerous clinical studies with various general immune system stimulating agents; the so-called chemoimmune therapeutics or immune stimulants. To this group belong bacterial preparations and extracts, their synthetic derivatives, fungal products and new chemical agents. They were all chosen for clinical testing because of their antigen-unspecific immune stimulating action and anti-tumour efficacy in experimental animals. The immunological stimulation, i.e. the enhanced activity of phagocytes, natural killer cells and other immune defence cells such as lymphocytes, could also be detected in patients, but the many controlled clinical trials so far have produced no evidence of a clear and reproducible anti-tumour action.

Some explanations may be proposed for the lack of anti-tumour action of chemoimmune therapeutics. There are for example, individual differences between patients, i.e. non-uniformity of clinical test groups compared with the uniformity of inbred mice in the tumour models, differences in the immunological reactions between spontaneously arising (autochthonous) human tumours and transplanted tumours, the defective or non-predictive methods for demonstrating appropriate immune defence, and insufficiently investigated dose-response relationships. Nevertheless, these arguments should not hide the fact that it can hardly be sensible at this time to test further chemoimmune agents in experimental tumour models with the objective of clinical tumour therapy, without discovering the cause of the poor predictive value of these tumour models. Elucidation is only possible by basic research into this specialised field, and this only by using animal experiments. A way out, should we wish to take it, is the tesing of chemoimmune therapeutics in animal models other than tumour models, with a potentially higher predictability for clinical work. For example, the chronic infection model seems to be far more clearly the result of a defectively functioning immune system than does neoplastic disease. However, the predictive power of these animal models of chronic infection for man must first be clarified by clinical testing of agents effective in animal experiments. If drugs of this kind also work clinically, we would have new ways of treating chronic, antibiotic-resistant infections which may frequently arise through neoplastic disease, or as a result of anti-tumour chemotherapy. With chemoimmune therapy there would thus be an indirect, if not direct anti-tumour effect. Essential requirements for achieving this goal, are the establishment of suitable animal models, and the testing of immune-stimulating agents on experimental animals.

As alternatives to antigen-unspecific tumour therapy with mediators or chemoimmune therapeutics, different possibilities for antigen-specific tumour therapy have been worked out preclinically. Of particular significance in this respect is the 'active' specific tumour therapy with combinations of tumour cells and immune system-stimulating substances (adjuvants) on the one hand, and the 'passive' specific tumour therapy using monoclonal (originally produced by one cell) antibodies obtained in cell culture on the other.

The foundations of active (so-called because the body is incited to react) specific tumour therapy were all formulated in animal models, i.e. in 'spontaneously' occurring chemical or virus-induced tumours, particularly in the mouse and rat. In my research for example, the concept of anti-tumour therapy using tumour cells mixed with the enzyme neuraminidase was devised. Combinations of tumour cells and neuraminidase, formulated in a precise immunization schedule, were clearly therapeutically effective in selected transplanted mouse tumours, as well as in spontaneously occurring mammary tumours in dogs. The prerequisite was that tumour cells from the disease-inducing tumour, or at least from tumours with related antigens (foreign matter) were used for immunization, were employed.

On the basis of these investigations, clinical trials show that the combination of tumour cells and neuraminidase has a distinctly therapeutic, i.e. life-prolonging, action in human colonic cancer. Certainly this finding must be verified before an antigen-specific tumour therapy for colonic cancer using tumour cells and neuraminidase can be generally recommended, but the data obtained to date indicate a new therapeutic potential for this particular neoplastic disease, largely unaffected by previous therapies. Within the scope of the animal research debate, it is useful to remember that this new form of therapy could not have been developed without animal research.

The use of monoclonal antibodies is the second form of antigen-specific tumour therapy. According to the technique developed by Köhler and Milstein (1975), antibody-synthesising lymphocytes, which do not survive in cell culture, are fused (hybridised) with mouse myeloma cells (tumour cells which, after hybridization with the lymphocytes, confer on them the capacity to survive in cell culture). The resulting hybrid cell produces an antibody, and those hybrids which form antibodies capable of binding to human tumour tissue and triggering consequent immunological reactions are then selected. Suitable antibodies are radioactively labelled and are tested for their suitability to specifically localise human tumours (xenotransplanted) in the nude mouse, using specific recognition methods such as radioimmuno-scintigraphy. Thus, the antibody is tested to determine whether it preferentially binds to the tumour after being given to a tumour-bearing nude mouse. If the test antibody is suitable for this purpose, similar clinical investigations are performed on patients, i.e. for the radioimmunoscintigraphic localization of tumours. Following positive tumour localization, a radioimmune therapy is attempted with an increased dose of the now radioactively-labelled antibody until the tumour's radiotoxic range is reached, or specific immunotherapy is performed by administering 'cold', i.e. non-radioactively-labelled antibodies. Clinical investigations of this kind are already being performed in various centres. The clinical results so far obtained, for example on ovarian cancer and melanoma, justify the hope that monoclonal antibodies offer new possibilities both for tumour diagnosis and for tumour therapy.

Whether the administration of antibodies alone, or that of combinations of antibodies with isotopes, toxins or cytostatics (so-called 'poisoned arrows' or 'magic bullets' created by combining cytotoxic agents) will be successful, only the future will tell. With particular reference to the development of 'poisoned arrows', numerous basic investigations, especially in animal models, still have to be performed. A detailed account appeared in the periodical series *Pharma-Dialog* 9 (1986).

References
Böckle, F. Das tier als gabe und aufgabe. Von der kreatürlichen verantwortung des menschen. In: Händel, U.M. Tierschutz: Testfall unserer menschlichkeit. S. Fisher, Frankfurt 1984.

Der Bundersminister für Jugend. Familie und Gesundheit: Tierversuche: Probleme und lösungsmöglichkeiten, Bonn 1982.

Deutsche Tierärzteschaft e. V.: Codex experiendi. Leitsätze für experimente mit tieren 1983.

Driscoll, J.S. The preclinical new drug research program of the National Cancer Institute. Cancer Treatm. Rep. 68, 1 (1984) 63-76.

Farber, E. The multistep nature of cancer development. Cancer Res. 44, 10 (1984) 4217-4223.

Fialkow, P.J. Clonal origin of human tumors. Ann. Rev. Med. 30 (1979) 135-143.

Fiebig, H.H., Henss, H., Schuchhardt, C. and Löhr, G.W. Xenotransplantation und chemotherapie menschlicher tumoren. Vergleich des ansprechens der nacktmaus. Verb. dtsch. Ges. inn. Med. 88 (1982) 966-969.

Frohner, E., Neumann-Kleinpaul, K. and Dobberstein, J. Lehrbuch der gerichtlischen tierheilkunde. Parey, Berlin 1955.

Galmeir, W.M. Onkologie: Ein kommentar zur situation in der praxis. Münch. med. Wschr. 125 (1983) 26, 613-615.

Goldin, A. and Venditti, J.M. Progress report on the screeening program at the division of cancer treatment. National Cancer Institute. Cancer Treatm. Rev. 7 (1980) 167-176.

Höffe, O. Ethische grenzen der tierversuche. In: Händel, U.M. Tierschutz: Testfall unserer menschlichkeit. G. Fisher, Frankfurt 1984.

Hoff, D.D., von Casper, J., Bradley, E., Sanbach, J., Jones, D. and Makuch, R. Association between human tumor colony-forming assay results and reponse of an individual patient's tumor to chemotherapy. Amer. J. Med. 70 (1981) 1027-1032.

Hoff, D.D., von Clark, G.M., Stogdill, B.J. and Sarosdy, M.F. Prospective clinical trial of a human tumor cloning system. Cancer Res. 43 (1983) 1926-1931.

Joss, R., Nöthiger, F., Greiner, R., Goldhirsch, A. and Brunner, K.W. Kolorektale karzinome. Fortschritte und ungelöste probleme. Schweiz. med. Wschr. 111 (1981) 697-705.

Kant-Brevier, hrsg. J. Pfeiffer, Schröder, Hamburg, 1974.

Khoury, T. Buddhismus. In: Brunner-Traut, E. Die fünf grossen weltreligionen. Herder, Freiberg 1984.

Köhler, G. and Milstein, C. Continuous cultures of fused cells secreting antibody of predefined specificity. Nature 256 (1975) 495-497.

Kraemer, H.P. and Sedlacek, H.H. A modified screening system to select new cytostatic drugs. Behring Inst. Min. 741 (1984) 301-328.

Loeper, E. von. Tierrechte und menschenpflichten. In: Händel, U.M. Tierschutz: Testfall unserer menschlichkeit. G. Fisher, Frankfurt 1984.

Lorz, A. Die entwicklung des deutschen tierschutzrechts. In: Händel, U.M. Tierschutz: Testfall unserer menschlichkeit. G. Fisher, Frankfurt 1984.

Max-Planck-Gesellschaft. Bericht und mitteilungen. Tierversuche in der forschung 1/1981.

Meyer-Abich, K.M. Das recht der tiere. Grundlagen für ein neues verhältnis zur natürlichen mitwelt. In: Händel, U.M. Tierschutz: Testfall unserer menschlichkeit. G. Fisher, Frankfurt 1984.

Narayanan. Alternative screening for cytostatics at the NCI. Bristol Myers Symposium, New Leads in Cancer Drug Research, London 1985.

Oeser, H. and Koeppe, P. Voraussichliche entwicklung der Krebssterblichkeit in der Bundesrepublik Deutschland. Münch. med. Wschr. 123 (1981) 706 ff.

Riecker, G. Ärztliche ethik und tierversuche. München. med. Wschr. 127 (1985) 134 ff.

Schopenhauer, A. Gesammelte. Werke. Hanser, München 1977.

Schweitzer, A. Die lehre der ehrfurcht vor dem leben. 2. Aufl. Union Verlag. Berlin 1963.

Sedlacek, H.H. How to find immunomodulators - a look backward and forward. Behring Inst. Mitt. 74 (1984) 122-131.

Sedlacek, H.H., Weidmann, E. and Seiler, F.R. Tumor immunotherapy using Vibrio cholerae neuraminidase (VCN). In: Jeljaszewicz, J. et al. Bacteria and Cancer. Academic Press, London 1982.

Sedlacek, H.H., Dickneite, G. and Schorlemmer, H.U. Präklinische prüfung von immunomodulatoren unter spezieller berücksichtigung von chemoimmuntherapeutika. Symposium "Entwicklung, Prüfung und Aswendung von biologisch aktiven substanzen in der Krebstherapie", Göttingen 1984.

Sedlacek, H.H. Tumorimmunologie und tumortherapie. Eine standortbestimmung. Beiträge zur Onkologie. Bd. 25. Karger, Basel 1987.

Senn, H,J. Indikationen und erfolgsaussichten der chemotherapie maligner tumoren. In: Brunner, K.W. and Nagel, G.A. Internistische Krebstherapi.e. Springer, Berlin 1979.

Sordat, B. and Wang, W.R. Human colorectal xenografts in nude mice: expression of malignancy. Behring Inst. Mitt. 74 (1984) 291-300.

Spaermann, R. Tierschutz und menschenwürde. In: Händel, U.M. Tierschutz: Testfall unserer menschlichkeit. G. Fisher, Frankfurt 1984.

Teutsch, G.M. Tierscutz als geschichte menschlichen versagens. In: Händel, U.M. Tierschutz: Testfall unserer menschlichkeit. G. Fisher, Frankfurt 1984.

Tierschutzgesetzblatt vom 24 July 1972, Bundesgesetzblatt Teil I, Nr. 74 vom 29.7.1972, S.1277. Dtsch. Tierärztebl. 9 (1972) 338.

Wasielewski, E. von and Sedlacek, H.H. Alte und neue probleme der tumorforschung. Dtsch. med. Wschr.

Wittke, G. Tierschutz aus ethischer, ethologischer und physiologischer sicht. Ther. d. Gegenwart, 115 (1976) 349-365.

Wolters, H.-G. Tierversuche auf dem prüfstand von Öffentlichkeit und politik. Tierversuch - experten sagen ihre meinung, abteilung für Öffentlichkeitsarbeit der Hoechst AG, Ernst Bäumler, Jürgen Fricke, S.7.

Woodruff, M.F.A. Cellular heterogeneity in tumours. Brit. J. Cancer 47 (1983) 589-594.

A bridgehead for alternative methods
Immunology
G. Wick

Immunology is the science which deals with human and animal defence systems. It is probably no exaggeration to say that this specialty, together with microbiology, has contributed most to the fact that the average life expectancy in highly developed countries now at about 75 years, while at the turn of the century in Europe it was only 50 years. In countries where modern understanding of immunology is limited or non-existent, the average life expectancy remains relatively low. In India for example, it is approximately 45 years. In the light of this, it should be appreciated that the immune defence system of man and animals does not operate independently, but is heavily dependent on nutrition, hormonal influences, and on communication with the nervous system. Thus, our own extended average life expectancy is not only due to preventive measures such as inoculation, and the resultant eradication of specific, previously fatal, infectious diseases, but also to improved hygiene and nutrition.

Immunology combined with genetics represents the most explosively growing research area of the last 20 years. This is demonstrated by the fact, amongst others, that about 1000 printed pages with an immunological content are published in the literature **per day**. Other parameters, for example the awards of Nobel Prizes for Medicine, which in recent years have been made in disproportionate numbers to immunologists, support this contention.

Structure and function of the immune system
The immune system has the task of protecting the integrity and the identity of an organism. Integrity is disturbed for example, by a bacterial, viral or parasitic infection, and identity is compromised by the introduction of a foreign body such as an organ transplant (i.e. non-identical genetic material). In both cases, the immune system strives to eliminate the foreign material, organisms or organs. This task can be achieved on the basis of three vital properties:

a) the capacity to differentiate between 'self' and 'non-self'
b) specificity
c) the ability to remember

In simple terms, this means that the immune system has the capacity to distinguish endogenous from exogenous (foreign) material and to react against the latter by eliminating it. The immune reaction is specific. Thus, vaccination or an infection caused by specific microorganisms (e.g. diphtheria bacilli) or their toxins, will protect an individual **only** against diphtheria and not against any other infectious disease (e.g. tetanus). The ability to remember refers to the generally known fact that the immune system 'remembers' encountering foreign matter, a long time after initial contact, be it active or passive. Thus, a suitably immunologically prepared organism can react to a second infection much more rapidly and effectively than it could to the first infection. This implies that, as is well known, one does not contract many infectious illnesses a second time. Thus in Vienna, after he caught and survived the plague, the fabled Augustin was able to nurse others without becoming reinfected with the disease, since he was protected by the ability to remember and the specificity of his immune system.

What exactly does the immune system consist of? It is not concentrated in any specific organ, but 'pervades' several organs, which for their part send out cells which circulate in the lymphatic fluid and the blood stream, and thereby reach all parts of the body. In particular, the bone marrow, thymus (thymus gland), lymph nodes and the spleen, all belong to the immune system, and which also consists of cells distributed throughout the body; some are stationary (e.g. in the skin), while others are mobile (certain white blood cells). These immune cells can be separated into different groups according to whether they take part in the uptake and recognition of foreign matter, in the subsequent transfer of recognition information to the cells, or in the defence process itself. Fig. 12 shows the most important organs which make up the immune system, while Fig. 13 is a diagrammatic representation of those processes which occur, from the active or passive penetration of foreign materials into the body, to their elimination.

All foreign materials which can result in an immune reaction in humans or animals, and which can therefore be recognised by the immune cells as 'non-self', are designated as **antigens**. When such antigenic material, e.g. a bacterium, penetrates an organism, it is first taken up by phagocytes (macrophages), taken apart, and presented on the surface in a form recognizable to other cells of the immune system. Certain white blood cells, the so-called T-helper cells, now recognise these antigens on the surface of the antigen-presenting cells, and act as distributors to those lymphocytes able to eliminate them (Fig. 13). These can either be the cytotoxic, or killer T cells, or the so-called B cells and their final forms, the plasma cells. In this example, the B cells, if they encounter further bacteria of the same species, undergo a differentiation and maturation process based on their capacity for recognition. With the support of the T-helper cells, they finally produce **antibodies**, which they then release into the blood and which become available throughout the body for defence against further infection of this kind (specificity!). Antibodies then attach to the surface of the bacteria, and by activating specific enzyme systems in the surrounding blood plasma, cause the disintegration of the bacterial membrane and thus destroy the bacteria itself.

In other cases the immune system does not use antibodies as the chief means of defence, but activates the **killer T cells** mentioned above. The difference between these two possibilities is that the killer T cells, albeit with the support of the T-helper cells, travel directly to the site of the infection. There, they come into close contact with the penetrating foreign matter ('fighting hand to hand') and destroy them. The organism particularly chooses this kind of defence in viral infections, in infections with bacteria which are difficult to penetrate (e.g. tuberculosis), and in genetically incompatible organ transplants etc. In these situations, the destruction of foreign cells and organs is not effected by antibodies, but by small molecular weight proteins (so-called cytotoxins) derived from the killer T cells.

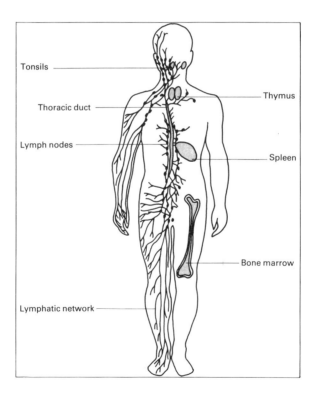

Figure 12. Diagrammatic representation of the most important tissues of the human immune system.

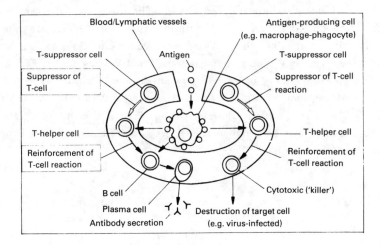

Figure 13. Diagrammatic representation of the progress of an immune reaction. (Shaded structure = blood and lymphatic vessels). Firstly, the foreign matter (e.g. bacteria) is taken up by phagocytes (macrophages) and presented on their surface (antigen-presenting cells). There can now be a response to contact with this material by antibody-producing cells (B-lymphocytes) or by cytotoxic lymphocytes (killer T cells). Both humoral and cellular immune reactions are subject to strengthening (by T-helper cells) and weakening (by T-suppressor cells) influences (Staines, Brostoff and James).

The T-helper cells play a central reinforcing role in the 'sociology of the immune system' (Fig. 13). A further important subpopulation of lymphocytes are the T-suppressor cells, which as their name suggests, exert a suppressive effect on the immune reaction. The strength of the immunological response to contact with an exogenous antigen depends on the subtly regulated balance between T-helper and T-suppressor cells (Fig. 13).

Immunopathology
So far only the positive, and thus protective aspects of immune reactions have been mentioned. However, disordered function of the immune system lies at the root of numerous diseases. To this group for example, belong the currently much publicized acquired immunodeficiency syndrome (AIDS), and also the various malignant tumours. Our bodies permanently produce tumour cells which are normally recognised and eliminated by immunological surveillance. The clinical occurrence of tumours has often been attributed to defective functioning of the immune system, which explains why patients undergoing immunosuppressive therapy, and children with congenital

immunodeficiency diseases, have a significantly higher incidence of tumours than normal subjects.

Of particular interest are those diseases in which the immune system is not beneficial, but is actually detrimental to the organism; the so-called **immunological hypersensitivity reactions.** Examples of these include allergies and various occupational eczemas, as well as the large groups of autoimmune diseases. The latter comprise diseases where the immune system paradoxically attacks and finally destroys endogenous cells, organs and organ systems. Many diseases of the hormone-secreting glands belong amongst the autoimmune conditions, e.g. juvenile insulin-dependent diabetes, as well as the various rheumatic syndromes, multiple sclerosis and many others. Finally, it should be mentioned that rejection of an organ transplant counts as an immunological hypersensitivity reaction, although the immune system is only acting normally to eliminate a foreign organ. Nevertheless, this process is associated with life-threatening complications for the affected patient.

Together with attempts to gain further insight into immune function, especially the mechanisms of immune regulation, modern immunology is preoccupied with the possibility of correcting hyper- or hypoimmunity by appropriate therapeutic measures. The potential for influencing the immune system in positive or negative ways is embodied in the concept of **immunomodulation.**

Immunomodulation measures for immune deficiency diseases include replacement of immunological components and products in which the patient is deficient. For example, bone marrow transplantation may be performed in humans following high dose irradiation, or protective antibodies isolated from the blood of previously immunised individuals may be injected (e.g. gamma-globulin injections for protection against hepatitis). Attempts are now being made to transfer not entire cell populations, e.g. bone marrow, but only specific mediators, i.e. those messenger substances by which cells of the immune system communicate with each other and by which they accomplish their defence function, e.g. the destruction of foreign cells. Such positive immunomodulatory molecules, some of which can now also be manufactured by gene technology, include interleukin 1, interleukin 2 and gamma-interferon. Most of these therapeutic approaches are still in the research stage, but together with the further development of gene technology, are of increasing practical significance.

Cases of immunological hypersensitivity in which the immune reaction should be suppressed, represent a much more difficult situation. The chief problem is that in patients with autoimmune diseases for example, the autoimmune reaction responsible for the destruction of an organ or an organ system, e.g. the joints, must be suppressed, while its full defensive capacity against other antigens, e.g. various micro-organisms, must be maintained. So far, this so-called antigen-specific immune suppression is in its infancy, and we have had to manage with non-specific methods, such as adreno-cortical hormones and similar agents such as hydrocortisone. These drugs not only influence the immune system and the abnormal immune reaction, but also many other body cells, and this frequently leads to undesirable side effects.

Methods supplementary to animal research
This book bears the title 'Alternatives to Animal Experiments'. There are no alternatives in immunology, and for some time to come, no alternatives can be foreseen. However, this does not mean that animal experiments cannot also be increasingly supplemented by other methods in this area of research. Previous sections have attempted to make it clear that the immune system consists of numerous different cells which are present in the most diverse organs, and which also circulate in the peripheral blood and communicate with each via a complicated network. Furthermore, the cells of the immune system have a close mutual relationship with other systems of the body, particularly the two other important communication systems, the nervous system and the hormone system. For example, the cells of the immune system are in close contact with the connective tissue system, e.g. the reticulin filaments, and it is therefore difficult to consider them in isolation. Thus, if one were to speak of 'alternatives to animal experiments', then it would only be on the supposition that certain preliminary investigations in cell culture could be performed. Nevertheless, their results would still have to be finally verified in experimental animals, and thus in the whole organism, before being applied to man. Furthermore it is important - as in all biomedical research - to remember that cell culture experiments or animal studies only give results relevant to man if that system mirrors the behaviour in man. Thus, it is difficult or impossible to breed tubercular bacilli in the rat, while it is very successful in guinea-pigs. Moreover, interestingly enough, one can only propagate the leprosy parasite in rather exotic armadillos, but not in various standard laboratory animals. Conversely, there are specific regulatory mechanisms of vital processes which remain constant from the simplest forms of life, e.g. bacteria, to man. This is the basis for the fact that most of the biomolecular knowledge derived from prokaryotes can also be applied to eukaryotes, and thus for human life processes. There are however, certain exceptions: modern gene technology frequently uses certain bacteria to express DNA sequences in mammals. However, many mammalian proteins function best if they have also undergone biochemical changes which first take place in the cells **after** the nucleic acid message has been translated. Such processes are termed **post-translational**, and include for example the attachment of sugars to amino acid chains (glycosylation). Since bacteria are not capable of glycosylation, some of the genetically-engineered prokaryotic proteins are less active than those produced from eukaryotes.

The scope of the foregoing account does not permit discussion of the numerous examples by which *in vitro* immunological methods can supplement *in vivo* investigations. These problems should therefore be discussed using selected examples from the research field of the Institute for General and Experimental Pathology of the Medical Faculty of Innsbruck University. In this regard it should be stressed that in the area of supplementing *in vivo* investigations with *in vitro* methods, immunology has assumed a pioneering role, from which many other disciplines are now also profiting.

Monoclonal antibodies
Everyone, even the most fervent opponents of animal research have, with or without their knowledge, received the benefits of the knowledge gained from animal experimentation. Here I am thinking not only of the development of vaccines and drugs, but also of those investigations which most people do not realise depend directly on animal research. For example, in most European countries, and so in Austria, every newborn baby is tested for various possible congenital defects (in Austria a total of about 10). These examinations deal with congenital diseases such as thyroid function disorders and metabolic diseases, in which the levels of certain substances in the blood must be detected. If such a disorder is present, appropriate counter-measures can often be taken. In some conditions, a life-long diet will prevent the most serious consequence of the disease, for example, total dementia. Determination of the blood levels of numerous substances, including drug levels in patients treated for hypertension, infectious diseases and epilepsy, and the determination of hormones in blood serum and urine, require very sensitive analytical methods, for which the detection range of conventional chemical methods is inadequate. Immunological methods are now widely used for this purpose. In principle, this works as follows: A rabbit for example is immunised with the human pregnancy hormone, human chorionic gonadotrophin (HCG), produced by the placenta. The animal recognises this hormone as foreign and produces antibodies against it. These antibodies can be isolated from the blood serum of the rabbit and, because of the specificity of the immunological reaction (see above), react only with HCG. If, under suitable experimental conditions, such a refined antibody is mixed in a test tube with the urine of a pregnant woman, an immunological reaction will occur, which allows the exact concentration of this hormone to be determined. All the commercially-available pregnancy testing kits work on this principle. The same applies to determination of blood and urine levels of other substances, some of which have already been mentioned.

In 1975, Cesar Milstein and Georges Köhler, who later won the Nobel Prize, succeeded in establishing a method for producing so-called monoclonal antibodies. These have revolutionised medical diagnosis, as well as the detection of various substances, in all areas of biological research. The principle of this method, using the example of producing antibodies against HCG, will be briefly outlined. As before, an animal, generally a mouse, is immunised with HCG. The immune system of this mouse will recognise HCG as foreign and will produce antibodies against it. Naturally, it is only possible to obtain a small volume of blood from a mouse, and so it would not be sensible to use such a small animal to produce antibodies for diagnostic purposes. The basis of the Köhler and Milstein method consists of removing the spleen from such a mouse, isolating from it those cells which are capable of forming antibodies against HCG, and fusing them with specific mouse tumour cells. The resultant structure, which is part spleen cell and part tumour cell, is known as a **hybridoma**. The spleen cell gives the hybridoma the ability to produce antibodies against HCG, while the tumour cell regulates its survival in cell culture. Complicated selection methods make it possible to choose, out of millions of hybridomas, the particular one which produces antibodies of exactly the specificity and binding strength required for a particular purpose. Since, by virtue of its origin, a hybridoma represents an identical cell family (a clone), the product of these hybridomas are known as monoclonal antibodies. The hybridoma now can be made to divide and grow in cell culture, providing us with a 'factory', which in suitable cell culture conditions and a sufficient time, can produce antibodies with the desired characteristics. This type of cell culture can be performed, depending on the amount of antibody required, in small containers, in large flasks and even in industrial fermenters. In principle therefore, although some mice must be sacrificed at the beginning, no further animal experiments will be necessary after the appropriate hybridoma has been established. Such hybridomas can also be frozen in liquid nitrogen at -196°C, stored for an almost indefinite length of time, and thawed and grown again when required.

HCG is produced in large quantities a few days after the start of pregnancy; a considerable proportion is excreted in urine. Detection of urinary HCG is thus a sure way of confirming pregnancy. Up until 20 years ago, a urine extract would have been injected into a mouse or toad. If it contained HCG, hormonal changes would be observed in the animal - a biological test for pregnancy (see p. 114). One laboratory which routinely carried out this pregnancy test used up to 10,000 mice or toads each year! Subsequently these tests were replaced with immunological methods based on polyclonal antibodies produced in rabbits. Now, pregnancy tests can performed with monoclonal antibodies, which can be produced exclusively in cell culture on a large scale.

Production of anti-inflammatory agents by genetic engineering
Certain adrenal cortex hormones, the so-called glucocorticoids, possess immunosuppressive and anti-inflammatory properties as well as their own specific hormonal action. These properties have been exploited, and glucocorticoid-like drugs such as cortisone have been successfully used to control inflammatory conditions. However, glucocorticoids exert an indirect, rather than a direct effect on those cells activated during inflammation, and which produce inflammation-promoting compounds such as prostaglandins. Indeed,

glucocorticoids are able to activate genes in phagocytes or macrophages which lead to the production of proteins which possess an anti-inflammatory action. The first such compound, which is induced by glucocorticoids and whose gene can be isolated, cloned and sequenced, is lipomodulin. This agent prevents prostaglandin production by inhibiting phospholipase A2, an important enzyme in prostaglandin synthesis. The action of cortisone is thus not direct but indirect, its effect being mediated by lipomodulin. Irrespective of their anti-inflammatory action, glucocorticoids also have specific side effects, which restrict their use as immuno- suppressive or anti-inflammatory agents. These include gastric ulcers, loss of bone substance and reduction of endogenous adrenal cortex function etc. Genetically-engineered lipomodulin and other, as yet undefined glucocorticoid-induced agents with similar properties would allow anti-inflammatory therapy without the undesirable side effects of the glucocorticoids.

Practically, this is done by treating macrophage cell cultures with glucocorticoids. Genes which were previously inactive, but which are activated by this treatment, are then isolated, cloned and sequenced from the DNA of the cells. These genes are then introduced into prokaryotic or eukaryotic cells using suitable vectors, which then express the appropriate proteins in large quantities and in pure form.

In principle, animal experiments are not needed to clone such genes. However, determination of the anti-inflammatory action of the genetically-engineered products still requires animal experiments. Inflammation is a complex process which occurs in the blood vessels, and which involves many other organ systems, particularly the connective tissue and nervous systems. Thus, a potential anti-inflammatory effect can only be tested in a whole organism.

Interaction between the immune and hormonal systems
As mentioned above, there is a 'dialogue' between the immune system and the hormonal system. As with every dialogue this is a two-way process. In the previous section it was mentioned that certain hormones can suppress the functions of the immune system. However, this does not only apply to adrenal cortex hormones, and it is now clear that other hormones, such as the catecholamines (from the supra-renal medulla and the brain), certain hormones from the pituitary gland (growth hormone, prolactin), and thyroid gland hormones may also influence the capacity for immunological reactions. Thus for example, the loss of the pituitary gland drastically reduces immunological defence, which can be restored by injections of growth hormone and prolactin.

Interestingly, lymphocytes also produce specific factors which influence the hormone system. Thus for example, adrenal cortex hormone levels, especially corticosterone and cortisol, rise following immunization with vaccine or infection. This probably accounts for the fact that the immune reaction does not continue interminably, but only produces enough antibodies to deal with the particular challenge. If cultures of adrenal cortex cells are mixed with the culture fluid of stimulated lymphocytes, an increased production of adrenal cortex hormones can be demonstrated. This system can also be used to test the effect of purified factors obtained from lymphocyte culture media, before they are used in experimental animals or humans. As mentioned previously, most cell culture work helps to provide specific new knowledge and to bring research to the stage at which experiments on the whole organism will then be necessary.

Aging research

The phenomenon of aging affects each individual and, quite apart from the socio-economic significance of this time of life, is one of our most powerful personal experiences. Nevertheless, by comparison with other areas, age research has assumed relatively little importance. Contributing to this is the fact that the individual often refuses to think about aging since he would otherwise find it more difficult to cope with the associated physical, spiritual and intellectual changes. However, this should not prevent society from paying more attention to this central problem of human existence. This applies as much to the purely somatic aspects of old age as to psychological, sociological and economic aspects. In western countries, the over-60 group already comprises 12-15% of the total population, and will increase still further in the next decade. The same is increasingly true, with some delaying factors, for the presently less highly developed countries, whose chief problem remains the high birthrate with lower average life expectancy.

Since aging is a multifactorial process, cell culture investigations are only partly representative of each process which occurs *in vivo*. However, certain questions can still be resolved in appropriately conceived *in vitro* studies.

For example, one interesting problem is the paradox that old people respond less well to immunological environmental stimuli (exogenous antigens) than the young, and are therefore more susceptible to infections and tumours, while they react more strongly to endogenous antigens (autoantigens). It is also a fact that the capacity of the immune system to differentiate between 'self' and 'non-self' becomes increasingly lost with age. One of the bases for altered immunological competence in age seems to lie in the fact that cellular cholesterol metabolism, including that of the immune system, no longer functions as well as it did in youth. The disorder arises from the fact that, amongst other things, lymphocytes and other body cells can no longer regulate their cholesterol uptake from the environment, and thus the intracellular production of cholesterol also suffers. This leads to a situation where the cell surface membrane becomes more viscous in old age, i.e. is 'stiffer' than in younger individuals. This makes the communication of lymphocytes with other cells, discussed above, more difficult and can lead to the phenomenon of altered immune reaction.

Lymphocytes from old people can now be isolated from the blood and specifically 'modulated' in cell culture, so that their surface membrane is made more fluid, i.e. 'rejuvenated'. This is possible by adding phospholipids (e.g. lecithin), and lymphocytes treated in this way have a similar function in cell culture to that of lymphocytes from young donors. This 'rejuvenation' can be reversed by introducing an appropriate form of cholesterol into the cell culture medium. Thus, after very extensive cell culture research, an improvement in immune reaction was finally achieved in living people (e.g. by research groups in Israel) by appropriate treatment with liquidised phospholipids.

Of course, the examples mentioned here do not include all possibilities for *in vitro* experiments to supplement and expand *in vivo* methods. However, they clearly show that in immunology, important and sometimes very surprising advances in this field have been recently been made. Thus, exploration of the causes, treatment strategies and possibilities for preventing acquired immunodeficiency syndrome (AIDS) is perhaps a good example of how the combination of bio-molecular methods, *in vitro* techniques and animal research have rapidly increased our knowledge about this new infection. The progress of research in molecular biology has resulted in the immunodeficiency virus, HIV (formerly HTLV/LAV), being quickly identified, and specific components of it can now be manufactured by gene technology. Using appropriate cell culture methods the virus can be propagated and used for diagnostic purposes, but ultimately, animal experiments will inevitably be necessary for any therapeutic or preventive progress. It is important to remember that AIDS research in experimental animals is only possible by using apes, particularly chimpanzees, since other animals cannot be infected with the AIDS virus and therefore cannot develop symptoms of the disease. Since it is already too late for a vaccine for infected people, and since the testing of an AIDS vaccine on healthy people is evidently out of the question on ethical grounds, there is no alternative to animal research.

Finally, there is an important role for computers, which, by accessing national and international databanks, not only prevent unnecessary animal experimentation, but also the duplication of experiments, be they *in vivo* or *in vitro*. Thus for example, if a gene is isolated and sequenced in a laboratory, then this sequence can, by access to an appropriate databank, be compared with all previously known gene sequences. This would then prevent someone in another laboratory from concentrating on an already known cloned gene.

In summary, it should be stressed that animal experiments, or the direct use of humans, are usually only used to confirm studies which have already been performed *in vitro*.

Protection of the unborn
Prenatal toxicology
D. Neubert

For parents it is undoubtedly a shattering experience for their child to be born with severe abnormalities. Often they will ask about possible causes, and they will wonder whether the abnormal prenatal development could have been prevented.

There are basically two causes of gross structural abnormalities:

1. 'endogenous', i.e. having a genetic or 'spontaneous' cause
2. 'exogenous', i.e. having an external cause

It is the doctor's problem to decide which mechanism has most probably led to the abnormal development in an individual case. This naturally presupposes that a definable cause can be found at all for the pathological development which has occurred. Only then does it become possible to prevent a similar abnormality in subsequent pregnancies.

It should also be stressed at this point, that, for the one-off case, it is almost never possible to pinpoint the cause of the abnormal development. Identification of an external factor as the causative agent of abnormal prenatal development is always based on statistical findings; the assessment of risk always remains a statistical prediction.

The incidence of 'endogenous' or 'exogenous' factors as causes of abnormal prenatal development
In man (and apparently in all primates), disordered prenatal development is not an infrequent occurrence. From our present state of knowledge we estimate that on average, almost 50% of all fertilised ova implanted in the uterus die before birth (about 35% soon after implantation, and 10-15% as clinically evident abortion). The reason for these prenatal losses should probably be sought in endogenous factors (e.g. 'spontaneously' occurring chromosomal abnormalities). The number of prenatal abnormalities is often much higher than that manifested at birth.

The proportion of exogenously caused embryotoxic and fetotoxic effects relative to the total number of abnormal prenatal developments is difficult (or impossible) to assess. It is possible that abnormalities due to a particular agent only stand out from the 'background' of predominantly endogenous abnormalities (about 3% of all live births demonstrate severe abnormalities) if they are caused in large numbers by a specific agent in a population (for example thalidomide).

At present, there is no strategy designed to reduce the 'spontaneous' rate of congenital (i.e. present at birth) abnormalities; indeed, attempts at intrauterine diagnosis of an abnormality may sometimes kill the developing child. With our present state of scientific knowledge we can only attempt to recognise the causes of exogenously occasioned congenital abnormalities and then to avoid them.

Detection or elimination of a prenatal toxic potential
Convincing medical and scientific data on the embryotoxic (or teratogenic) potency of an agent can only be obtained after extensive and meticulous human studies (epidemiology). It necessarily emerges from this that a harmful effect must first have appeared in humans before it can (hopefully at an early stage) be recognised. In any case hundreds, perhaps thousands, of abnormalities may occur before the cause of the damage can be clearly identified and the action of the pathogenic agent eliminated.

Although the importance of terato-epidemiologic studies is undisputed, the weight of evidence from such studies is often overvalued - especially by laymen. In principle, each single study can only arouse a *suspicion*; a causal connection can never be *proved*. A reliable picture (both for the detection of a specific effect and particularly for ruling it out) is obtained primarily after several carefully conducted and comprehensive (and correspondingly expensive) studies.

Epidemiological investigations on humans are therefore not usually suitable for primary prevention. The use of injurious agents can only be avoided if their toxic potential was already known at an early stage; i.e. before human exposure to them. This only comes about through appropriate experimental investigation; here lies the value of such studies.

Current approaches to testing for reproduction-toxic effects
Testing for potentially toxic effects of a substance on reproduction is now exclusively performed by experiments on intact mammals. Toxic effects on reproductive behaviour in mammals can appear before fertilisation (impairment of fertility, germ cell mutations) or during prenatal development. Corresponding effects are thus detectable either before or at birth, or are first manifested postnatally (sometimes even relatively late in life). Because of the multiplicity of possible abnormalities and sites of action appropriate 'models' are now used to evaluate the expected reproduction-toxic potential of a substance. These models facilitate the assessment:

- either of the development of mammalian organisms over two or three generations (two or multi-generational studies),
- or of different stages of development in several studies (e.g. phase I, II, and III trials).

These experimental arrangements are now used to study the current reproduction-toxic potentials of an agent (or a physical factor such as radiation).

Possible alternatives to testing for reproduction-toxic effects
If consideration is given to how current strategies and assay methods used to assess possible toxic effects on reproduction can be modified or fundamentally altered, it must be established why such an alteration of the existing situation is sought. The underlying motives differ considerably, and may be influenced on scientific, 'ethical', political and 'economic' grounds.

It is certainly reasonable to discuss appropriate *scientific* grounds. Ultimately, there is the question of whether existing methods could be improved or even replaced by other techniques. Naturally, for this to be achieved, strong evidence and improved risk evaluation with human relevance is required.

The methods to be discussed in the field of reproduction-toxicology have been divided into two categories for practical purposes:

a) *Primary* testing (primary stage testing), i.e. methods used when there is no relevant toxicological information available about the substance in question.
b) *Secondary* testing (secondary stage testing) i.e. to elicit further information about substances about which some toxicological data already exists.

To date, because of the complexity of the question (*excluding* as many toxic effects as possible), biologically complex 'primary stage testing' methods have been used. Not surprisingly, these are mainly methods which use the intact mammalian organism as a 'model'.

Conversely, as far as 'secondary stage testing' is concerned, *in vitro* methods have become extremely important in reproduction toxicology in the last 15 years. These methods make it possible to select and implement techniques (e.g. cell cultures) which are very specialised, simplified and purpose-designed, and are aimed at specialised and well-defined areas of research.

Relevant *political* or *'ethical'* grounds cannot be discussed rationally from the scientific standpoint. It is ultimately a question of whether the current methods are socially acceptable and of the safety standards required. However, this can present a problem if, for example, a high standard for drug safety or environmental health is demanded, but animal experiments are seen as unacceptable for the acquisition of relevant toxicological information.

Economic requirements (safety should cost less) also play a role in an age in which, even in industrialised countries, research funding is always limited. Here too there are inconsistencies, if further demands (e.g. for the wholesale substitution of *in vitro* methods for animal experiments) are made in such a situation.

In some areas of toxicology it should be quite possible to achieve such a dual aim (e.g. in testing for potential carcinogenicity). However, in other areas of toxicology, attempts at drastic reduction of experiments on intact animals will be confronted with the surprise that batteries of *in vitro* tests will cost rather more than corresponding *in vivo* tests.

In vitro methods which can act as substitutes in reproduction toxicology
To date, there is no way in which mammalian development, from preimplantation to the fetal phase, can be investigated *in vitro*. At the present state of knowledge, it seems improbable that such a technique will become available in the next decade.

There are two possible approaches to the production of 'models' for prenatal development, which would perhaps prove suitable for routine toxicological investigations:

1. Culture systems with mammalian embryos or organ rudiments from mammalian embryos, in which *partial* aspects and specific *phases* of prenatal development can be investigated.

2. Culture systems with *lower* organisms, in which the majority of the development can be observed.

Using the first approach, it is unlikely that *one* culture system can provide all the information required in 'primary stage testing'. Instead, *batteries* of tests of suitable systems are needed (if indeed relevant systems could and should be used for toxicological evaluation of this nature!) It is conceivable that a combination of a 'whole embryo' culture could be amalgamated with certain organ culture systems for this purpose.

So far, no suitable system and no groups of culture systems have been *validated* for this project (i.e. for 'primary stage testing'). In this context, validation refers to the comparison of the new method with the existing *in vivo* routine investigations (e.g. phase I-III tests).

Even if it should be shown that it is possible to discover the *teratogenic* potency of an agent with such *in vitro* methods, other prenatal toxic effects (e.g. the induction of congenital dysfunctions), which are currently recognisable with the usual phase I- and III-trials (*in vivo*), would remain undiscovered with such *in vitro* studies.

It cannot yet be judged whether the use of the second culture approach, in which most of the test species development can be observed, is sensible and will generally produce the evidence needed for reproduction-toxicological investigations. To date, whether such 'models' have any relevance for the routine determination of toxic effects in mammalian organisms, has not been established. Corresponding *in vivo* - *in vitro* comparisons and validations must be performed over the next decade in order to evaluate the possible importance of such test systems. Even so, at best only possible *teratogenic* effects would be assessed, and the same limitations would apply to the other potential reproduction-toxic effects mentioned above.

References

Neubert, D. Prospektive in-vitro-modelle als ersatz für langzeituntersuchungen. In: Zur problematik von chronischen toxitätsprüfungen. AMI-Berichte. Reimer, Berlin 1 (1980) 227-287.

Neubert, D. In-vitro models in pre-natal toxicology. Europ. Teratol. Soc. 13th Conf. Sept. 1985, Rostock Teratology, 33 (1986) 4A.

Neubert, D. The use of culture techniques in studies on pre-natal toxicity. Pharmac. Ther. 18 (1982) 397-434.

Neubert, D. Benefits and limits of model systems in developmental biology and toxicology (in-vitro techniques). In: M. Marois: Prevention of physical and mental congenital defects, Part A: The scope of the problem. Liss, New York 1985 (p. 91-96).

Neubert, D. Toxicity studies with cellular models of differentiation. Xenobiotica 15 (1985) 649-660.

Neubert, D. Misinterpretation of results and creation of 'artifacts' in studies on developmental toxicity using systems simpler than in-vivo systems. In: J.W. Lash and L. Saxen: Developmental mechanisms, normal and abnormal. Progress in Clinical and Biological Res. vol 171, Liss, New York 1985 (p. 241-266).

Neubert, D. Results of in-vivo and in-vitro studies for assessing prenatal toxicity. Environment. Health Perpect. 70 (1987) 89-103.

Reward and punishment
Behaviour research and ethopharmacology
F. Lembeck

> There is no creature too small,
> That man cannot learn from him.
> The same magic is all around us
> Near and far,
> And the world remains full of mystery -
> From the flea to the stars.

Observations on animals

It is usually very difficult to understand what Nobel prize winners write. However, there are exceptions: In 1940, Karl von Frisch wrote a small book entitled: 'Ten tiny household companions'. In it he described the *modus vivendi* of the fly, the midge, the flea, the bug, the louse and other creatures so enchantingly that we feel an empathy with them. With the books 'He talked to the cattle, the birds and the fish' and 'How man went to the dogs', Konrad Lorenz gave a charming insight into the behaviour patterns of all kinds of creatures. Both scientists received Nobel prizes for Physiology and Medicine in 1973; Karl von Frisch for having explained a mechanism for the way in which bees communicate, and Konrad Lorenz for having discovered the mechanism of imprinting by which a lifelong bond between young and old animals (or their human substitutes) is forged. Both scientists provided considerable knowledge for the social behaviour of man.

Charles Darwin strikingly showed the similarities between the emotional reactions of humans and animal (Figure 14). We know only too well from everyday observations, that the muscles of the human face express the feelings of the heart. Body movements are equally expressive, as every good actor demonstrates when, at a distance from the stage we only see his movements but not his facial expression. Experienced psychiatrists often recognise a patient's mental state from his appearance and bearing, movements and expression, before he opens his mouth (Dixon, 1986). The term 'ethology' refers to the scientific concept of behaviour and the habits of humans or animals in their natural surroundings, as well as the developments giving rise to them.

It was natural that one should also wish to study the physiological functions of this behaviour under the influence of drugs. Wilhelm Busch (Figure 15) presented early findings of this kind. One of the unexpected scientific discoveries resulting from animal behaviour studies is described below:

J.F.J. Cade, a psychiatrist in a small Australian hospital with no research experience, suspected that disorders such as depression or mania could be caused by excessive production of endogenous substances. Perhaps he had read the work of Sigmund Freud, who had similar suspicions. He injected (how awful!) urine from patients and healthy subjects into the abdomen of guinea-pigs, because he believed that these substances would be excreted in the urine. He then tested the effect of uric acid, since it was present in great amounts in the urine. Because it is almost insoluble he injected the lithium salts of the uric acid and, as controls for a possible lithium action, he gave other animals lithium carbonate. After two hours, the latter animals became lethargic and had forgotten their nervousness, although they did not sleep. They recovered after several hours. He then administered the lithium salts - without ethical or other approval - to a patient who had been hospitalized for five years with chronic mania. He gave lithium salts over several days and found a marked improvement after five days. After two months of treatment the patient left hospital and resumed his work. Today lithium is - after comprehensive experimental and clinical safety investigations - the agent of choice in the treatment of mania (Tosteson, 1981).

Figure 14. Charles Darwin striking described the great similarity in expressing emotions (anxiety, excitement, fright) in man and animal with these two drawings (Charles Darwin, in the translation by J.V. Carus, The Expression of Emotions in Humans and Animals, Schweizebart'sche Verlagshandlung, Stuttgart 1872).

He croaks delightedly
And has to stand on one leg.

Figure 15. After taking liqueur, Hans Huckebein's raven is completely drunk and demonstrates distinctly human behaviour (Wilhelm Busch). Behaviour research on rats under controlled laboratory conditions provide many inferences for man. Learning and memory ability can be measured, social behaviour can be analyzed, the therapeutic action profile of substances with psychotropic effects can be quantified and differentiated. Ten years of observation by Jane Goodall in a colony of free-living chimpanzees in Gombe Stream National Park in Tanzania, teaches us what historically developed inheritance man conceals in his being - including the tendency to commit gruesome acts against the nearest members of his own kind.

Watching a colony of monkeys is usually the biggest attraction for children on a visit to a zoo. They see that young monkeys get up to more pranks than they themselves! Observing gorillas and chimpanzees in their natural habitat in the jungle is rather more difficult and dangerous. The American Diane Fossey lived in a small hut in the midst of a tribe of mountain gorillas for 18 years before she was murdered by poachers. The Englishwoman Jane Goodall has lived with a tribe of chimpanzees for 25 years. Both scientists were tolerated as members of the tribes and had uninterrupted insight into all aspects of the lives of these animals; high academic recognition was accorded to the work of these researchers (Fossey, 1985, Ghiglier, 1986). Birth and death, love and trust, organised attacks, murder and cannibalism were observed amongst these animals. How human...!

Research

A major high point in animal behaviour research was probably the well known studies of J. Pavlov when he showed how animals are able to learn, and that this is expressed not only in their behaviour, but also in their 'conditioned' (i.e. acquired) autonomic reflexes. Later, Skinner in particular devised special tests in which rats could be induced to perform or to avoid certain actions by 'reward and punishment' reactions. The conditioned learning and forgetting reactions of rats was used by de Wied to study memory capability. Following the discovery of the first psychotropic drugs in 1950, many of these experimental systems became very important in the pharmacological search for new agents of this type. Thus, it became possible not only to observe, but also to measure movement coordination, muscle rigidity, reaction to stimulus (e.g. blowing with an airstream) and other functions. This resulted in the highly active psychotropic drugs used in modern psychiatry. One test compound was apparently an exception; even in high doses it led to none of the effects recognisable in these tests. Indeed rats treated with it seemed to become particularly docile. When studies were repeated at a zoo, aggressive or nervous animals also demonstrated this calming and settling effect. This was the discovery of Valium from the benzodiazepine group; drugs which, if they are not taken indiscriminately, are valuable and relatively harmless tranquillisers.

Modern ethopharmacology encompasses not only the behaviour of the individual animal but also its social behaviour with its own kind. It differs in this from other branches of pharmacology in which the mechanisms of action of substances on the nervous system are obtained by biochemical or electrophysiological methods (Dixon, 1982). A prerequisite for the direct observation of experimental animal behaviour is to maintain them in optimal conditions. Mice, rats and hamsters sleep throughout the day. Thus, they are nocturnal animals, and as long as there is no light, they drink, find food and move about in their cages. Their metabolic processes and hormonal secretions are also adapted to this day-night rhythm. All these functions must be taken into consideration in ethopharmacology.

If two rats, which have previously been in isolation, are put together in a cage, they behave similarly to two strange people who meet in a railway compartment. One, a gentleman, who wants to impress with much noisy unfolding of a large newspaper, and the other, a lady, who thereupon looks out of the window and turns her back to him which gives him a good opportunity to admire her figure. These 'elements' of behaviour can be well defined in research animals, categorised into functional groupings and measured. Aggressive and depressive, sexually influenced, inquisitive or anxious components can therefore be measured by observation.

Thus, as the mental wellbeing of the human depends on organised behaviour amongst his peers, so it is with the mouse. If they have enough space, they reserve a personal corner which they defend. The emission of odorous substances in their urine prevents the female or young animal from being attacked (Dixon et al., 1984). If mice are attacked they take up a defensive stance; they expect to be hurt and their nervous systems activate centres which lead to a higher tolerance to pain.

The sensitivity of these reactions can also be demonstrated by biochemical means. If a rat is taken out of its familiar cage and put alone in a 1 m^2, lighted and empty box, this unfamiliar environment leads to a brain-regulated secretion of the 'stress hormone' ACTH. The rat probably feels as we would if we were put down naked in the middle of a football field! If a mouse, after having been kept by itself for a long time, is put with a group of other mice in a communal cage, a conflict occurs between the 'intruder' and the other mice, measurable by recording the individual elements of their behaviour.

The use of ethopharmacological investigations can avoid the customary methods used for training animals; for example rewarding (with food) when the animal 'jumps', and threatening with punishment (showing it the whip) if it remains sitting. Only the behaviour elements are studied. This opens up a very subtle means of evaluating psychotropic drugs. Their administration is the only problem, since holding the animal and giving it an injection has already affected it. The test drug is therefore put in the drinking water. With this method for example, the effect of small doses of Valium can be seen, which was not possible with previous screening methods.

The brains of different species of animal and of humans differ considerably in their relative sizes and in their intellectual capacities. However, the electrophysiological and neurochemical processes in the brain are the same overall. The behaviour elements are also broadly similar, and ethological observations and ethopharmacological investigations (i.e. drug effects measured by ethological methods) enable extrapolation of their effects to humans.

Is ethopharmacology an alternative?
There is certainly no alternative method in which the living animal can be completely replaced by the use of isolated organs or cell cultures. Ethopharmacology rests on research into the behaviour and reactions of small laboratory animals such as mice, rats and hamsters. From this, an analytic appreciation of the individual elements of their behaviour is derived. These experiments are performed on non-anaesthetized animals, which are exposed to 'upsets' comparable to those which humans encounter daily. This represents a supplement or alternative to other methods, in which reflex or motor reactions to various stimuli (e.g. temperature, acoustic, optical, electrical stimulation) are used.

Ethopharmacology facilitates the study of psychotropic drug effects, and considerably supplements electrophysiological and neurochemical data for the molecular basis of their action. Ethopharmacology also possesses good predictive value for the action of psychotropic drugs on man, and is therefore one of the most elegant alternative methods.

References

Dixon, A.K. Ethopharmakologie: ein neuer weg zur untersuchung des einflusses von medikamenten auf das verhalten. Triangel 21 (1982) 95-105.

Dixon, A.K. Ethological aspects of psychiatry. Arch. Siss. Neurol. Psych. 137 (1986) 151-163.

Dixon, A.K., Huber, C. and Kaesermann, F. Urinary odours as a source of indirect drug effects on the behaviour of male mice. Ethopharmacol. Aggression Res., Liss, New York 1984.

Fossey, D. Mein leben für die affen. Geo. 5 (1986) 124-136.

Ghiglieri, M.P. The social ecology of chimpanzees. Scientific American 252 (1985) 84-93.

Frisch, K. Zehn kleine hausgenossen. Mohr, Tubingen 1947.

Lorenz, K. So kam der mensch auf den hund. Borotha-Schoeler, 1954.

Lorenz, K. Er redete mit dem vich, den vögeln und den fischen. Borotha-Schoeler, 1949.

Tosteson, D.C. Lithium and mania. Scientific American 244 (1981) 130-137.

Brain research
The nervous system under investigation
R. Gamse

The nervous system is an organ of the body which developed at a very early stage in evolution. Its function can be simply viewed as taking information from its surroundings or from the body itself, interpreting this information and ultimately contributing to the body's subsequent reactions (Fig. 16). The component parts of the peripheral and central nervous system are similar in principle, having nerve cells with processes of different lengths which make contact with other nerve cells. Transmission of information is achieved at these junctures by chemical messenger substances (neurotransmitters). Research in this field over the last 10-20 years has markedly increased our knowledge of these substances. During this time, conclusions were abandoned which were regarded almost as dogma in the past decades. Thus, the principle had been established that a nerve cell only utilises one neurotransmitter. However, it now seems much more likely to be the rule rather than the exception that several neurotransmitters facilitate information transfer from one cell to the others. This naturally increases the potential for 'fine tuning' of information transmission.

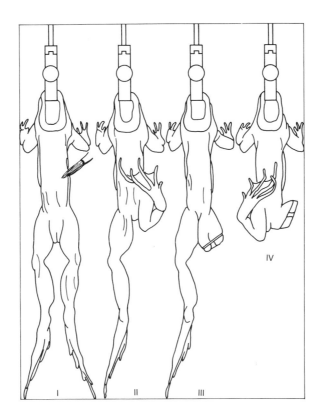

He croaks delightedly
And has to stand on one leg.

Figure 16. *Spinal cord decision making:* The picture shows the scraping reflex after clamping with no involvement of the head. I. The right rear side was dabbed with acetic acid using a paint brush. II. A scraping reaction by the right lower limb followed which removed the acetic acid. III. The right leg was amputated. IV. After again irritating the *right* rear side, a scraping movement was performed with the *left* hind le.g.

This coordinated defensive reaction has been recognised by physiologists for over 100 years. It was previously thought that the spinal cord only consisted of the conductor channels and the distribution points. The observation of the scraping reflex, being a significant coordinated reaction, gave rise to extensive discussion, since coordinated functions had long been attributed only to the brain. This single test represented a decisive contribution to research into the functions of the central nervous system. Through this, the adaptability of the nervous system was demonstrated experimentally for the first time. It would hardly have been possible for modern clinical neurology to develop without this basic knowledge (M. Verworn: *Physiologisches Praktikum.* G. Fischer, Jena 1932).

Study of brain functions in animal experiments
The areas of brain research are just as numerous as the functions of the nervous system. These functions include the sensory functions, regulation of circulatory and respiratory activities, involvement of the endocrine glands and thus also of transmission and regulation of sleep/waking behaviour to name just a few. These functions have the same basic stages of development in animals as in man, and results obtained in animal research can therefore be directly extrapolated to humans. Methods used in animal experiments embrace histological, surgical, biochemical and electrophysiological investigations, which are variously combined with behavioural tests. In the past, the anatomical interaction of nerve pathways was studied by surgical intervention. Today, selective neurotoxins which only block specific types of nerve cell may be used for this purpose. Thus, after application to specific areas of the brain, both the anatomical course of the affected nerve pathways and the effect on the animals' bodily functions or behaviour can be studied. While functional defects can be determined by these methods, functions of individual parts of the brain can be activated by stimulation using chronically implanted electrodes. In electrical stimulation tests the researcher obtains information about the localization of centres, but not about the neurotransmitters which participate in these effects. To obtain this information, additional agents which influence the action of a specific transmitter are frequently injected into the brain. If the stimulatory effect is thereby altered, a role for that particular compound is indicated. However substances injected into the brain can also trigger behaviour changes in which spontaneous brain activity is modified. They can depress or activate experimental animals, inhibit or enhance muscle activity and induce or deter sleep etc. By using these neuroactive substances in brain research, animal models can also be developed for human diseases. Thus, the therapeutic actions of a new drug can be deduced from its efficacy on the animal model (Sontag, 1981).

In vitro methods of brain research
The costs of whole animal tests are high, and alternatives have therefore been sought. The *in vitro* methods currently in use will be presented in the subsequent text. However, the integrated function of the central nervous system means that these methods are only suitable for the study of relatively small component functions. They can therefore only be substituted in specific cases for investigations of the whole animal, but may frequently supplement them. The first step in the transition from *in vivo* to *in vitro* experiments employs brain sections obtained from animals which have been painlessly killed. Since the supply of oxygen from the nutrient solution limits the size of the section, only local control loops and influences of afferent nerve endings can be examined. The method is mainly employed for studies of neurotransmitter release and their effect on electrical processes of the nerve cells, although inactivation processes such as breakdown or cellular uptake can also be studied. Isolated nerve endings, obtained by homogenisation of brain areas and subsequent separation methods, also lend themselves to these problems. New

drugs for treating depression for example may also be tested with this *in vitro* method. Homogenisation of nervous tissue also leads to preparations of nerve cell membranes, in which the binding sites (receptors) for particular neurotransmitters may be examined. The advantage of this method lies in the fact that relatively little tissue is needed. Thus, about 1000 determinations may stem from one preparation of a single rat brain. Consequently, in the screening processes (see p. 27) during drug development, numerous compounds can be tested and the ineffective ones 'filtered out' before being tried in animal tests. This method of testing has therefore contributed considerably to the reduction in the use of experimental animals which has taken place in recent years.

The use of cell culture in brain research is currently increasing since the discovery of culture conditions which allow the sensitive nerve cells to survive. For some time it has also been possible to cultivate nerve cells from the nervous system of growing animals. Thus, viable nerve cells can now be obtained during surgery, and possible changes occurring in diseased animals can be studied. The characteristic property of nerve cells, i.e. that they do not proliferate in growing organisms, does not preclude the use of cell culture, although all experiments must be carried out on the limited number of cells of the original cultured tissue. Cell culture is primarily used to study the development of cell processes and their neurotransmitters, for the development of cell contact sites and for electrical processes induced by the transmitters. However, it may also be suitable for testing toxic chemical effects on the nervous system. One of the ultimate aims is the development of therapeutic methods for injuries to the nervous system, for example quadriplegia, a presently incurable condition.

From experimental research to therapy
The use of methods in developing forms of therapy for diseases of the nervous system is multi-faceted. The causes of these diseases are largely unknown. With two kinds of condition, for example Parkinson's disease and Alzheimer's disease, nerve tracts in specific areas of the brain are basically involved. This leads to functional deficiencies, which manifest themselves in the observed clinical symptoms. With other diseases such as endogenous depression or schizophrenia, no cellular failure is detectable. Thus, purely functional disturbances, ie, over- or under-activity of individual nerve tracts, must be accepted as the causes for these disorders. The extent to which alternative methods are employed in this area of brain research, and how far research on whole animals or patients should be supplemented, is discussed below.

Parkinson's syndrome (Morbus Parkinson) was first described in 1817. The discovery that Parkinson sufferers had a massive loss of nerve cells in one region of the brain stem, did not affect treatment until the 1960's. However, this situation has now changed as newly developed histological and biochemical methods have helped to identify the neurotransmitter contained in these destroyed nerves. It thus seemed possible that treatment of Parkinson's disease could be achieved by administering the missing agent. After animal experiments had shown that preliminary administration of the neurotransmitter dopamine, lacking in Parkinson's disease, could reverse an experimentally-induced dopamine deficiency, the first administrations to Parkinson patients showed similar clinical improvements. This therapeutic principle of substitution, which had already been used successfully in endocrinology, was revolutionary in brain research. Today it represents a mainstay in the treatment of Parkinson's disease. Since the missing transmitter has been identified, *in vitro* methods for developing new drugs for Parkinsons disease are being extensively introduced. The receptor sites for the neurotransmitters may be identified using appropriate nerve cell membrane preparations. This lead to the development of drugs which would exert the same effect as dopamine at these sites. The breakdown of dopamine could be further elucidated by biochemical methods. Thus, it would be possible to use only *in vitro* methods to search for compounds which could inhibit this degradation. Indeed, one of these agents is already used in treatment. Although all these new drugs were initially examined by *in vitro* methods, their therapeutic efficacy on animal models for Parkinson's disease had to be established. The extrapolation of the observed effect to the patient was limited, so that in all animal models, only partial aspects of the disease could be studied. In 1983, symptoms of Parkinsonism were observed by chance in young drug addicts. Careful studies subsequently revealed that the injected drugs contained a contaminant which destroyed the same nerve pathways as in Parkinson's disease. Just like Parkinsonism patients, the affected drug addicts found less relief after a short time from therapy with the anti-Parkinson treatments currently available. The compound also induces Parkinson symptoms in experimental animals. Thus, experimental brain research into Parkinson's disease now seems to have found an optimal animal model. Potential new therapies which possess few side effects and which have long-lasting results can now be tested in this model. Furthermore, previously unknown factors which lead to the onset of Parkinson's disease can be studied. Thus, the development of agents which hinder this onset might also be possible. If one considers that 1% of the over-65 year olds suffer from Parkinson's disease, then the significance of this research for the improvement of quality of life of an ever-aging population may be evaluated.

While with Parkinsonism, disorders of mobility are predominant, in Alzheimer's disease the intellectual faculties of the patient are reduced to imbecility. The search for treatment is therefore geared to an area of specific human brain functions. The use of behavioural tests in animal models in which 'memory capacity' is trained principally by reward or punishment, is therefore only conditionally possible. Similar difficulties also arise with the development of therapy for schizophrenia, a derangement condition, which can lead both to states of delusion and complete apathy. In animal models, only drugs which eliminate the delusional states can be tested (Mathysse, 1983). Nevertheless, the introduction of these drugs has revolutionised the treatment of schizophrenia.

References

Hodos, W. Animal welfare considerations in neuroscience research. Ann. N.Y. Acad. Sci 406 (1983) 119-127.

Langston, J.W. In: S. Fahn: Recent developments in Parkinson's disease. Raven Press, New York 1986.

Mathysse, S. Making animal models relevant to psychiatry. Ann. N.Y. Acad. Sci. 406 (1983) 133-139.

Sontag, K.-H. Tierversuche - Eine methode bei der entwicklung zentralnervös wirkender arzneimittel. Edito Cantor, Aulendorf 1981.

No pain for the animal?
Pain research
R. Gamse

Pain is an integral part of human life. The sensation of various types of pain is the most frequent reason why patients seek out a doctor. It is therefore understandable that pain relief has always been an important aim of the healing art. Various methods, from the administration of drugs to hypnosis are used to this end. This demonstrates that the phenomenon of pain not only contains a sensory component, but that psychological components also influence pain intensity (Fig. 17). In endogenous depression these can be so strongly overpowering that patients experience pain even though no peripheral pain stimulation could be detected. It is also known that the threat of pain can enhance or attenuate its perceived intensity. Numerous drugs with a range of pharmacological actions are available commercially. The question is therefore whether the development of new preparations through pain research is either necessary or ethically defensible.

Acute and chronic pain

Pain can be divided into acute and chronic forms. Acute pain is an important warning signal to life, which signals possible tissue damage due to chemical, thermal or mechanical stimuli. This form of pain also gives early warning of illness.

In spite of extensive research, the physiological bases of the perception of acute pain can be only partially explained. The transmitters of the nerve tracts, which conduct pain stimuli from the spinal cord to the cerebrum are unknown. Acute pain can usually be eliminated in most cases by the availability of existing drugs. Side effects, such as damage to the gastric mucosa and bone marrow, or addiction, which can occur after taking strong analgesics of the opiate group, justify the search for better drugs.

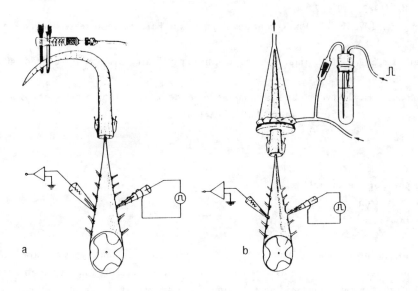

Figure 17. *Pain research in the isolated spinal cord.* The cone-shaped lower part of each picture shows the lower section of the spinal cord of a rat with the nerves emanating from it. Electrodes for nerve stimulation and conduction of action currents are applied to these nerves. In addition, the nerves from the rat's tail (above) lead to the spinal cord. Tail and spinal cord were taken from a newborn rat; the entire preparation has a length of 2 cm. In the left-hand picture (a) the pain nerves of the rat tail are stimulated by pressure with a clamp, in the right-hand picture (b) by soaking the rat's tail with a substance which penetrates the skin and reaches the pain nerves. This method was used to determine the type of nerves through which pain is signalled, and then to show that morphine or endogenous morphine-type peptides inhibit the transmission of pain impulses to the brain in the region of the spinal cord.

Experiments with these techniques last for months, and the electronic outlay is considerable. However, it is recognised that alternative methods which are able to give clear cut results are often very complicated (Yanagisawa and Otsuka).

In contrast to acute pain, chronic pain has no warning function, but represents an agonising suffering for the afflicted person. The pain frequently forms part of an underlying syndrome which must be treated. Examples of this type of pain include pain in cancer patients, back pain, stump pain after limb amputation and frequent headaches and neuralgia, e.g. nerve pains associated with herpes zoster. The causes of such pain are still insufficiently understood. In many cases, drug therapy which alleviates acute pain is unsuccessful for chronic pain. This makes the patient switch from doctor to doctor and from clinic to clinic, resulting in numerous therapeutic efforts and repeated operations. It is therefore apparent that studies into the causes of these pain forms, and the development of effective modes of treatment is urgently needed.

Investigations on the conscious animal
A major part of pain research is performed by animal experimentation. Since sensitivity to pain in man is influenced by many subjective factors, the extent to which results of animal experiments may be applied to humans can be questioned. In numerous pain research studies, pain is deliberately inflicted on the test animal. They therefore run counter to ethical guidelines that pain in animal experiments must be avoided. All tests on animals which involve pain or suffering for the animal are subject to specific authorization, and the granting of permission is legally controlled. Consequently, such experiments may only be carried out if they directly serve the needs of man and animal, or of scientific knowledge, and if the aim of the experiment cannot be achieved by other methods. The legislative authorities lay down the ethical guidelines for animal experimentation into pain research in cooperation with scientific bodies and publishers of specialist scientific journals. Thus, in the journal *Pain* for example, only results which have been obtained under strict observation of these guidelines may be published.

It is not just in experiments into pain research that animals suffer pain. Table 3 shows the degree of pain severity in a variety of animal experiments and their frequency. The degree of severity ranges from the brief pain of a needle prick to the pain of experimentally-induced joint inflammation lasting several weeks. In both industry and universities, over 90% of all animal experiments are associated with slight to medium-grade pain for the animal. As can also be seen from the Table, pain research in animals should not be equated with excruciatingly painful animal experiments. Category 1 (medium grade pain) includes a large proportion of experiments into pain research, i.e. all tests in which evasion reactions are observed. In these studies, pain stimulation is applied to the animal until it responds with a motor reflex or a behaviour reaction, i.e. avoids the stimulus, or until an upper time limit is reached. This is chosen so that no injury occurs to tissue at the site of application. Therefore the tests are repeatable and the effect of a drug on the same experimental animal can be examined. By virtue of the fact that in all these pain studies, the experimental animal identifies the end of the pain stimulus by a reflex response or behaviour reaction, the duration of the influence of the stimulus is confined to seconds, the possible sensitivity to pain to tenths of seconds. To some extent, these tests are similar to a dentist testing the reaction of a tooth with dry ice. The most frequently investigated avoidance reactions should be mentioned briefly. In the tail-flick test, a heat-pain stimulus is applied to the mouse or rat tail. In the hot-plate test, experimental animals are placed on a 50-55°C hot plate and the time taken to begin licking the paws is recorded. In tests on mechanical pain stimuli, the mouse tail is pinched with defined intensity, or increasing pressure is exerted on the paw. With the 'shock-titration test', electrical stimulation is applied to sites on the skin of the experimental animal. The strength of the stimulus is increased until the trained animal signals the sensation of pain by pressing a button to reduce the intensity. This test allows continuous measurement of the pain threshold and is applicable to pain research in rats, primates and humans. The effects of drugs in these tests also correlate with their action in acute pain in patients, i.e. the test results are widely applicable to man.

Table 3. Degree of severity of pain in animal experiments

Category	Degree of intensity	Experiment	% of experimental animals		
0	slight	Injection, blood sampling, short-term immobilization, organ removal with fatal anaesthesia, surgery under anaesthetic with final death	68*	82**	61***
1	medium	Surgical operation under anaesthesia with regained consciousness and slight resultant injury, avoidance reactions	26	10	
2	high	Operations with considerable pain suffering or damage which require and necessitate analgesia	2	7	38
3	very high	Operations with considerable pain suffering or damage which use no analgesia	4	1	1

* Animal experiments in the German pharmaceutical industry
** Animal experiments in Swiss universities
*** Report of the University of Washington

The causes of chronic pain and their possible treatment cannot be studied by the above-mentioned tests. For this purpose, inflammatory pain models are now predominantly used. One method consists of injecting carrageen into the rat's paw, which after a few hours leads to local inflammation. Another method involves the application of dead bacteria to cause joint inflammation in rats resembling that seen in rheumatic patients. The analgesic and anti-inflammatory effects of anti-rheumatic agents can be assessed by both these methods. The pain induced by these experiments is of considerably higher intensity (Category 3) and duration than that of avoidance reactions, and the aim must therefore be to replace them by other less painful animal experiments, or by methods which are performed without experimental animals.

Alternative methods
Since pain is an experience of higher life forms, its study in man is the only alternative to animal research if all aspects of pain are to be investigated. Partial aspects can be however be studied on anaesthetized or decerebrated animals. An anaesthetic inhibits sensitivity to pain, and to a certain extent, other reactions to painful stimuli as well. This disadvantage does not arise after decerebration of the experimental animal, i.e. after transecting the pain pathways to the cerebrum. The animals are therefore pain-free, but the mechanisms and modulation of pain conduction in the spinal cord can be studied by electrophysiological methods. These experiments, which are still performed on the whole animal, thus require expenditure on the appropriate equipment. However, they are an alternative to experiments on the conscious animal, with the advantage that the only pain inflicted on the animal is from the injection of a short-term anaesthetic.

In vitro models similar to those used in brain research are also used in pain research. The initial material is always tissue from painlessly killed animals. Amongst these *in vitro* tests are studies on tissue sections, in which the release and effect of neurotransmitters, as well as their pharmacological influence, can be examined. Experiments with nerve cell membrane preparations are intensively used for the investigation of new opiate-type painkillers. However, these experiments are only possible because the neurophysiological bases for the analgesic actions of opiates are known. Analgesics with other modes of action cannot be found in this way. This also applies to the use of cell culture in pain research. It will therefore be possible for these methods to replace a considerable number of studies on the whole animal when our knowledge of the origins of pain and pain conduction is greater.

Human studies
Clinical trials into pain research using volunteer test subjects and patients are now being increasingly performed. Different methods now permit the quantification of pain and thus

aid statistical processing of the results. Nerve impulses following pain stimulation can be detected by microelectrodes, and the measurement of electrical currents in the brain enables the recording of stimulations which are triggered by pain stimuli in the cerebrum. In this way, more objective measurements can be added to the data derived from test subjects. The pain stimuli used are similar in both animals and man, i.e. mechanical, thermal or electrical stimuli. This form of pain research in man thus highlights the problem that the mechanics primarily of acute, but not of chronic pain, are explained.

References
Editorial: Ethical standards for investigations of experimental pain in animals. Pain 9 (1980) 141-143.
Tierversuche in der Deutschen Pharmaindustrie. Erhebung des Bundesverbandes der Pharmazeutischen Industrie 1984.
Tierversuche an Schweizer Hochschulen. Neue zuricher zeitung vom 2./.3. März 1985.
Report of the University of Washington to the Department of Agriculture. Seattle 1982.

Animals help animals
Animal experiments improve the maintenance, feeding and treatment of domestic and useful animals
D.F. Sharman and Margarethe Holzbauer-Sharman

Purring happily, our cat sits on the desk and watches me working. We do not have to worry about the animal. She thrives on obviously delicious commercial cat food; the protein, carbohydrate, fat, vitamin and mineral content of which is compiled scientifically. Timely inoculation against cat 'flu and enteritis (panleukaemia) protect her against the most lethal diseases. Regular dosing with anti-helminthics controls the worm infestations which are unavoidable in rural cats which hunt mice. Ectoparasites are successfully combatted by treating the fur with insect powder. Injuries, usually as a result of nocturnal hunting expeditions, are dealt with by the veterinary surgeon under anaesthetic, and heal uneventfully after antibiotic treatment.

For a domestic pet to enjoy this carefree existence (Figure 18), numerous physiological and pharmacological investigations were necessary, leading to the understanding of the functions and requirements of the animal organism (pure research), as well as to the development of drugs and vaccines. The importance of animal experiments has already been emphasised in many of the preceding sections. The development of veterinary medicine has paralleled human medicine in many areas. In a way, since the reliability of drugs for patients is only achieved after favourable results in human clinical trials, man may sometimes be viewed as an 'experimental rabbit' for the animal.

Figure 18. A "dog's life": a well-known method of keeping dogs out of their owners' beds (approved by the Dogs Protection Society, the Society against the Misuse of Dogs, and by dogs themselves), is to set up an extra bed exclusively for canine use. To make this a welcome invention for dogs, it must have certain peculiarities. A slight movement of the head makes food available to the dozing dog. The feather mattress, not too soft, adapts to the shape of his body. Steps up to the bed save him from having to jump. Cushions can be used for reclining as well as for chewing. To lull the dog to sleep a music centre is installed. Favourite pin-ups facilitate sweet dreams. A large rotating fan keeps him cool without disturbing his sleep (from S.Baker, E.Gurney: *Living with Neurotic Dogs*, Heyne Books 503).

The great number of species-specific diseases creates an additional problem in veterinary medicine. In order to establish the causes of these animal diseases, and to find a successful treatment, research must be aimed at the appropriate animal species. This raises the problems of maintaining useful animals, for example, the questions of optimal housing conditions.

Because of the diversity of these veterinary problems, it is only possible in this short article to present some typical examples to show how animal experiments can lead to the improvement of living conditions for animals. The most recent research results were chosen as examples. The search for alternatives to animal experiments is also considered.

Contagious animal diseases and vaccines
Protection against infectious diseases by inoculation is one of the most successful areas of medicine (see also p. 153). Indeed, it has been possible to almost completely eradicate many contagious diseases by prophylactic vaccination. In the long chain between the disease, isolation of the pathogen, development of an active vaccine and successful inoculation of the person or animal at risk, animal research is indispensable. For passive immunisation, hyperimmune sera were obtained, for example, from the serum of animals which had been inoculated with the diluted or dead pathogen and whose immune systems subsequently developed specific antibodies. In future it may be possible, by cloning immune cells which produce specific antibodies in culture, to create immune sera independently of living animals. Immunologists have already succeeded in analyzing the amino acid composition of certain antibodies and synthesising the antibody *in vitro*. Some examples are given below, which illustrate the effect of new immunological discoveries on animal medicine.

1. Feline leukaemia
Feline leukaemia is a wide-spread malignant disease caused by a retrovirus (FeLV). It is a lymphosarcoma rather than a true leukaemia, and in particular, is a deficiency of the immune system (FAIDS, feline acquired immunodeficiency syndrome). Immunisation against a virus which impairs the immune system is a particularly difficult problem. Nevertheless, R.G. Olson (University of Oslo), in collaboration with the pharmaceutical company Smith Kline, has recently succeeded in producing a vaccine against FeLV on a new basis. This vaccine consists of two proteins: a virus antigen and a tumour antigen. It protects against the immunosuppressive action of the virus and also against the lymphosarcoma and other tumours occasioned by the virus. It is hoped that the development of this vaccine will lead to valuable collaborative studies in cancer and AIDS research.

2. Enteritis in young calves
This frequently occurring disease is mainly caused by rotavirus and K99-*Escherichia coli*. Recently, scientists at the Moredun Research Institute of the Agricultural and Food Research Council (U.K.) succeeded in developing a combined vaccine against these pathogens, which inactivates rotaviruses and *E. coli*. Since the calf will only get the best

protection if the appropriate antibody enters the stomach and intestinal tract orally, the pregnant cow is vaccinated. Antibodies against the pathogens are subsequently excreted into the colostrum and the milk, and are therefore drunk by the sucking calves during the first weeks of life.

If, for unavoidable reasons, the colostrum is not drunk, then the calf can also be protected against various pathogenic strains of *E.coli* by an oral vaccine which has been developed in the veterinary faculty of Munich University. With this vaccine, the young animals must be given large amounts of heat-treated pathogens with their food over a period of at least 10 days.

In rare cases in which the disease unavoidably occurs, the drinking of large volumes of salt- and sugar containing water can protect against dehydration and thus against death. This important observation was first made when treating humans during cholera epidemics.

3. Goat and sheep pox

In some Third World countries, goats and sheep are an important food source for wide sections of the population, and the control of goat and sheep pox is therefore very important. Investigations in England in 1982 have shown that a goat, which has previously had goat pox, also becomes immune to sheep pox. After isolating these pox viruses it was possible to develop a vaccine which protects both species of animal against pox. Field trials, e.g. in Yemen, have demonstrated the practical value of this vaccine.

4. Foot and mouth disease

In many countries, foot and mouth disease is still combatted by slaughtering the affected animals and all those animals which have been in contact with them. It is hoped, however, that this drastic remedy can be replaced in future by the introduction of an economically acceptable prophylaxis against this disease. The development of vaccines against foot and mouth disease is therefore regarded as being of the greatest importance. Thus, in the Animal Virus Research Institute in Pirbright (U.K.), with international cooperation, a 'foot and mouth disease vaccine bank' has recently been established, in which 500,000 injections of concentrated 'killed viral antigen' are stored in liquid nitrogen. The expected active life of vaccines stored in this way is 10-15 years, whereas previously, these expensive vaccines have only remained active for one year. A further immunological advance was made with the chemical synthesis of proteins of the foot and mouth disease virus. Injection of this pure antigen stimulates the formation of neutralising antibodies against the active virus in the hide (DiMarchi *et al.*, 1986).

5. Tuberculosis
With the development of the tuberculin test it became possible to recognise and separate those cattle suffering from tuberculosis. This disease has consequently been largely eradicated in many countries. Thus, in England in 1930, 20% of cattle gave a positive tuberculin test, but by 1963, this number had dropped to 0.07% (Ritchie, 1963).

The history of the development and significance of vaccines for protection against contagious animal diseases which, particularly in Africa, leads to considerable suffering and death in many herds of animals, was collated by Professor Wooldridge (1947). Among others, these contagions include rinderpest and anthrax. On the other hand, the prevention and treatment of rabies needs still more exhaustive studies.

Disorders of mineral metabolism
Deficiency in certain salts can lead to dramatic symptoms in animals. Grass tetany and milk fever are examples of diseases in which successful treatment was only achieved by pure research on animals to determine the normal concentrations of salts in the blood.

Grass tetany occurs in cattle and other grazing animals with an extremely acute course. The animals suddenly stop grazing, become overexcited, foam at the mouth and have muscle tremors. Some other external disorders lead to senseless galloping and bellowing. Equilibrium disorders occur and the animals suffer from cramps. If the animal is not helped within 30-60 minutes it dies. Careful studies of the blood of sick animals show that grass tetany is associated with an acute drop in blood magnesium level. Sometimes there is also a lesser drop in the blood calcium concentration. The consequence of this discovery was a simple but strikingly effective treatment: intravenous injection of a mixture of magnesium and calcium salts leads almost directly to recovery and to continued grazing for the animals. The switch from winter fodder, in which the magnesium content of the grass is low, in the feeding of young animals is often responsible for electrolyte disturbance in grass tetany. This is particularly so after spraying with potassium- and nitrogen-rich artificial fertilisers, since this reduces the magnesium uptake from the soil.

Milk fever (parturient paralysis) is another mineral metabolism disease, which occurs in many species of animals when young are born. Generally, the symptoms are muscle weakness, circulatory collapse, disturbances of consciousness and, especially with horses, cramps. The cause of the disease is now known to be a drop in the blood calcium content, which is probably brought about by increased calcium excretion in the colostrum and in milk. Calcium uptake from the stomach is also reduced. Treatment is effected by parenteral administration of calcium salts. Some grazing areas, as in Austria, contain very few calcium salts. Calcium salts are regularly added to animal fodder in these regions.

(For details and further mineral metabolism - and vitamin deficiency - disorders, see Blood et al., 1979).

Pharmacology
Many useful drugs can be taken from human medicine directly into veterinary medicine. However, additional problems are created in animal medicine by the species-specific metabolism of different drugs in the body, which can only be resolved by experiments on animals. Furthermore, to treat the manifold diseases which occur in the animal kingdom, specific drugs must be found and tested on the appropriate animal species. The first example drawn on here from animal pharmacology illustrates a method for developing anti-inflammatory drugs for horses and ponies. Later, problems of parasite control and animal anaesthesia will be discussed. Two further examples show the contribution of pharmacology to controlling the maintenance of useful animals and to the 'species conservation' of wild animals.

1. Anti-inflammatory agents
An 'animal-friendly' method for testing anti-inflammatory agents for horses and ponies was distinguished by the Ciba-Geigy Prize for 'Research into Animal Health'. Muscle and joint inflammatory conditions are not uncommon in these species and can lead to severe deformities. The new method consisted of implanting polypropyl sponges (5.0 x 2.5 x 30.5 cm) soaked in carrageenan, a substance which causes inflammation, under the skin of the neck. Inflammation then occurs in the surrounding connective tissue, and the exudate is absorbed by the sponge. The sponge is then removed, and the exudate collected and analyzed. As well as lymphocytes, these exudates contain arachidonic acid metabolites and some prostaglandins, typical indicators of an inflammatory process. When the anti-inflammatory substance phenylbutazone (Intrazone, Arnolds Veterinary Products, Reading, U.K.) was injected into the jugular vein shortly before the administration of carrageenan, no prostaglandins were found in the exudate, an indication of the anti-inflammatory action of phenylbutazone. This research model is extremely valuable for the development of anti-inflammatory agents for ponies and horses. Furthermore, the animals do not seem to find these investigations disturbing, since their behaviour and food intake remains normal (Higgins et al., 1984).

2. Parasite control using 'natural' compounds
The tremendous advance in the control of bacterial infections since the discovery of penicillin is now well known. Penicillin is a 'natural' substance, which is produced by the fungus, *Penicillium notatum*. The most recently discovered avermectines, produced by the soil bacterium, *Streptomyces avermictilis*, are also natural agents. Researchers from the pharmaceutical company Merck, Sharp and Dohme found that a double avermectine derivative, Ivermectin, has no antibacterial effects, but displays overt activity against

animal and human parasites. Horses, cattle, sheep and pigs can also be protected by Ivermectin against attack by worms, insects, mites and ticks. Thus, parasite control previously practised with dips, powders or sprays, to which toxic organophosphates or chlorinated hydrocarbons had been added, is replaced by the administration of small amounts of Ivermectin. Endoparasites too, such as the stomach and intestinal worm *Ostertagia*, which leads to uncontrollable diarrhoea in young cattle, can be rendered harmless by timely injection of a single small dose of Ivermectin at the larval stage. Indeed, Ivermectin is effective against most parasites in the larval stage. GABA, a neurotransmitter present in the peripheral nervous system of these life forms, is probably stimulated, which leads to their paralysis. Ivermectin is also important in human medicine where it is used for the treatment of onchocercosis. This is a widespread disease in West Africa and Central America, caused by a worm which lives in the skin for years, and which produces microfilariae which end up in the eye. A single Ivermectin tablet sees the disappearance of the microfilariae in the skin within a month. For a complete cure, the therapy must be repeated after 6-12 months.

3. Animal anaesthesia

The anaesthetic agents used in veterinary medicine are often the same as in human medicine. Methods of anaesthesia differ, however, in certain animal species. For anaesthetizing reptiles, immersion in 10% ethyl alcohol, which is absorbed through the skin, is suitable. The addition of the anaesthetic agent to the water is also used for fish where it is absorbed through the gills. The most frequently used anaesthetic agent is tricainmnethanesulphonate (MS 222). For anaesthetizing goldfish, ether is also added to the water (10-15 ml/l). Goldfish can also be anaesthetized by being embedded in crushed ice. It is not advisable to handle anaesthetized fish since they easily go into shock. With birds, special care must be taken to maintain their body temperature under anaesthesia. For anaesthetizing mammals, the steroid anaesthetic agent alphaxolone (Saffan; Glaxo, U.K.), is recommended since it can be given intramuscularly or intravenously (for details of animal anaesthesia, Green, 1979).

4. Sedative agents (tranquillisers)

This group of drugs is used to sedate the animals, but keep them conscious. This is desirable, for example, to perform smaller operations under local anaesthesia. The efficacy of tranquillisers is very different with different species of animals. Horses, cattle, sheep, dogs or cats are sedated by the injection of xylazine; they allow themselves to be handled and small operations may be performed. On the other hand, xylazine has almost no effect on pigs. The only tranquilliser which has so far been used successfully on pigs is stresnil. The markedly aggressive behaviour which can lead to severe injuries amongst piglets from different litters can be mitigated by the administration of Suicalm. Giving this drug can also help in cases where the mother pig bites and treads on her own young.

5. Pharmacological immobilisation of animals

The need to immobilise an animal 'pharmacologically' can arise in situations where, for example, a wild animal has to be captured. For this purpose etorphin (Immobilon, M 99), a synthetic morphine derivative, is particularly good. In many species, the action of this substance is considerably higher than that of morphine. Thus in humans, it is 200 times more effective than morphine, but in the hippopotamus, it is 80,000 times more potent. Etorphin can be shot at the animal to be captured with a gun. The captured animal becomes immediately unconscious and can then be transported over long distances without problem. After injection of diprenorphin (Revivon) animals are on their legs again within a few minutes. This method was used in Africa to round up of herds of wild animals, when wide areas of their natural grazing lands became flooded during the building of the Kariba Dam. Both drugs have therefore made an important contribution to the conservation of many species of wild animals. Injection of etorphin can also be used to immobilise large animals in zoos if, for example, it should be necessary to perform surgical operations.

As a matter of interest, it should be mentioned that etorphin is ineffective on sea-lions. Sea-lions can be immobilised with a combination of ketamine and diazepam. (for further details on animal anaesthesia and pharmacological immobilisation see Green, 1979).

Abnormal behaviour patterns

It is accepted that ever since man began to bring animals into his own circumscribed environment, abnormal, stereotypic behaviour patterns have appeared in these animals. These can be observed in open captivity, in an animal pen or in zoos. These stereotypes are mostly attributed to inadequate environmental conditions, insufficient space or unattractive surroundings. In modern zoos, in which animal enclosures have been widened and the animals' natural habitat imitated, stereotypes are now more rarely seen. Sometimes, apparently slight changes in the environment can normalise behaviour. For example, the repetitious wandering back and forth of anteaters stopped when a 20 cm layer of sand was spread on the ground. In livestock maintenance, behaviour disturbances present an ethical and economic problem, which behaviour researchers have been called upon to study. Some of these disturbances are so deep-seated that they can lead to self-injuries. For example, the ways in which piglets scrape their snouts on the ground or wall, or incessantly bite food troughs or their own tails, are painful and costly behavioural disturbances. A study of this produced a simple solution since the tail-biting problem always remains in pigs which are kept in straw. Early 'weaning' of piglets also leads to abnormal behaviour patterns and is now no longer recommended. The use of tranquillisers to control animal behaviour has already been mentioned.

Prevention of stress in the maintenance of useful animals
In Great Britain the maintenance of livestock is under the control of the Ministry of Agriculture. In 1968 the Ministry issued a list of regulations with which animal breeders must comply. The first paragraph states:

"Any person who causes unnecessary pain or unnecessary distress to any livestock for the time being situated on agricultural land and under his control, or permits any such livestock to suffer any such pain or distress of which he knows or may reasonably be expected to know, shall be guilty of an offence under this section".

In modern livestock maintenance, owners are particularly enjoined to prevent the occurrence of stress. Although we all think we know what is understood by 'stress', no simple definition of this concept has yet been produced (see for example Levine, 1985). The organism responds to 'stress' with a series of physiological and pathological reactions which are subsumed under the expression 'adaptation syndrome' (Selye, 1946). The extent of these reactions depends on the conditions in which the organism found itself when confronted with the stress situation. Reactions to acute 'stress' may differ from those of chronic 'stress'.

'Stress' leads in particular to an increase in hormone secretion from pituitary gland (e.g. vasopressin, oxytocin, prolactin, adrenocorticotropic hormone), adrenal cortex (corticosteroids) and adrenal medulla (adrenalin). In order to understand the connection between 'stress' and hormone secretion, numerous experiments to measure the blood and urinary concentrations of these substances in rats and rabbits, as well as humans, were necessary.

Animals kept in chronic stress conditions may show abnormal growth, reproduction and behaviour, as well as reduced protein synthesis and decreased production of immunoglobulins (see Landi *et al.*, 1985) with all their ensuing consequences. Different 'stress indicators' are used to recognise such situations in animal maintenance. A practical example from poultry raising is the continuous radiotelemetric recording of heart rate which is accelerated by the stress-conditioned elevation of adrenalin levels.

So-called 'choice experiments' can also help to determine whether a given situation is considered unwanted by the animal. In such studies, the animal can choose its environmental conditions itself (Ingram and Mount, 1975). For example pigs, sheep or calves can learn to press electric switches or to block an infrared light source with their snouts. In this way electric lamps, radiators, ventilators or water taps can be switched on or off. Such experiments have shown that young piglets often switch radiators on when the temperature of their surroundings is 20°C, while this only seldom occurs at a temperature of 25°C. On average, in a 24 hour period, calves and sheep choose to have more than 15 hours of light in their quarters. Ventilators are almost always shut off by pigs at temperatures above 27°C. The results of these experiments can be very useful in the design of stalls and for establishing different conditions for animal maintenance.

However, it should not be forgotten that the animal, just like man, does not always choose those conditions which are best for his wellbeing. Pigs for example tend to overeat if unlimited food is at their disposal. Nature normally regulates the wild animal's food supply.

An example of a practical stress situation often facing livestock is transport in motor lorries. Transport can even lead to sudden death in particularly sensitive animals, and the quality of the meat can also be impaired. It is therefore very important to minimise the stress components for animals in transit.

To assess transportation stress objectively, a 'transport simulator' was built which comprised an uncovered wooden container in which pigs, sheep or calves could be kept (Stephens et al., 1985). The apparatus was driven by a motor which moves and shakes the container in all directions, creating noise comparable to that of street traffic. Inside the container there is a push-button which the animals touch with their snouts to temporarily stop the vibration and noise. This is learned particularly quickly by pigs and is done as often as possible; a sign that the vibration is undesirable. Noise alone is less often shut off. Blood samples can be taken from the animals to determine various hormone levels. Thus it was found that vibration and noise together lead in pigs and calves to a raised blood corticosteroid content.

These examples show the importance of animal experiments for the control of diseases in domestic animals, for the development of new veterinary drugs and for proper animal maintenance. Experiments on animals, in which alternative methods are also substituted for specific areas of research, are unconditionally necessary for further improvements.

References

Blood, D.C., Henderson, J.A. and Radostits, O.M. Veterinary Medicine. A textbook of the diseases of cattle, sheep, pigs and horses. Balliere Tindall, London 1979.

DiMarchi, R., Brooke, G., Gale, C., Cracknell, V., Doel, T. and Mowat, N. Protection of cattle against foot-and-mouth disease by a synthetic peptide. Science 232 (1986) Reports 639-641.

Green, C.J. Animal anaesthesia. Laboratory Animals Ltd., London 1979.

Higgins, A.J., Lees, P. and Taylor, J.B. Influence of phenylbutazone on eicosanoid levels in equine acute inflammatory exudate. The Cornell Veterinarian 74 (1984) 198-207.

Ingram, D.L. and Mount L.E. Man and animals in hot environments. Springer, Berlin 1975.

Landi, M., Kreider, J.W., Lang, C.M. and Bullock, L.P. Effect of shipping on the immune function of mice. In: Archibald, J., Ditchfield, J. and Rowsell, H.C. The contribution of laboratory animal science to the welfare of man and animals. G. Fischer, Stuttgart 1985 (p. 11-18).

Levine, S. A definition of stress? In: Moberg, G.P. Animal stress. American Physiological Society, Bethseda, Md. 1985 (p. 51-69).

Ritchie, J. The eradication of bovine tuberculosis and its importance to man and beast. Stephen Paget Memorial Lecture (Research Defense Society) London 1963.

Selye, H. The general adaptation sydrome and the diseases of adaptation. J. Clin. Endocrin. 6 (1946) 118-127.

Stephens, D.B., Bailey, K.J., Sharman, D.F. and Ingram D.L. An analysis of some behavioural effects of the noise and vibration components of transport in pigs. Quart. J. Exp. Physiol. 70 (1985) 211-217.

Wooldridge, G.H. What animals owe to experimental research. Stephen Paget Memorial Lecture (Research Defense Society) London 1947.

We thank Dr. D.B. Stephens, The Royal Veterinary College, University of London for advice on questions of veterinary medicine.

Animals teach students
Animal experiments in lectures and in practice
F. Lembeck

"The clinical professor has to teach at the sickbed, or with the help of pathogenic material, the professor of pathological anatomy with cadavers, and the professor of experimental pathology with experiments".

"I should also like to stress here that it would be impossible to train the prospective doctor as an experimental researcher. Theatres are not there to mould whole sections of the population into actors, but rather to bring poetry to the audience".

"To clarify the contrast between old style general pathology and experimental pathology as I teach it, I recall the words of former doctors, according to which, black cataract is an illness where the patient cannot see, and neither can the doctor. Up to 1873, the status of general pathology in Austria may be characterised similarly. The professor saw nothing from the pertinent results and neither did the students". (Stricker 1896, 1897).

Heart and brain, arteries and veins, muscles and nerves have long been known in their anatomical form. However, the function of these organs was only first established in the 18th and 19th centuries. It began with 'vivisection' which refers to the opening of the body of an animal paralysed with curare, in order to expose the organs and *observe* their function. At that time, there were no anaesthetic methods available, even for humans. These organ functions were first recognised by observation, but not with quantitative recording equipment. Rokitansky, the Viennese Professor of pathological anatomy, saw more clearly than anyone the need to supplement morphological findings with experimental investigations. Stricker, the first Viennese experimental pathologist, defined the tasks of his Faculty with the moving words quoted above. Attempts were made to demonstrate the students' bodily functions through experiments. He constructed the forerunner of our modern projection equipment to demonstrate arterial blood flow to students.

The visible demonstration of functions in a teaching setting fell to a number of leading physiologists. The high point of this development was probably the lecturing of Tschermak-Seysenegg, a physiologist in Prague from 1913 to 1938. In his lectures, one animal experiment led to another, and there may have been over 200 experiments performed each year. For the students this was very impressive, although a great burden for the assistants, but it was generally a very versatile form of instruction. Experiments in lectures were soon complemented by practical classes, in which students carried out simple experiments themselves, mainly on frogs, although more complicated experiments were demonstrated by the assistants. New clinical methods were introduced, such as haematological diagnosis, blood pressure measurement and ophthalmoscopes which students could use on their colleagues. The development of modern clinical methods of investigation enables many experiments in modern physiological practice to be performed by students on colleagues. In one English institute, I observed how medical students introduced stomach tubes into each other to study gastric juice secretion. The students commented very factually that it may even be worthwhile to have the experience of a patient whose stomach tube is introduced by a relatively inexperienced colleague. Drug actions can also be presented in a pharmacological practice session. Here only the teaching value of such an experiment is significant.

It would be pointless to demonstrate the lowering of blood sugar levels by therapeutic insulin administration in rabbits, since this can equally well be represented in slides. The student ultimately sees this effect in clinical teaching on patients. However, the injection of a large insulin dose to a mouse, which leads to hypoglycaemic cramps and unconsciousness can be important, since from this the student learns how carefully insulin must be administered. With this type of experiment, it can be shown that glucose injection immediately eliminates this 'insulin shock'. Such a test is simple and of considerable relevance to clinical practice. Only through observation and contact can muscle relaxation under anaesthesia, muscular rigidity due to neuroleptics or agitation due to stimulants be learned graphically. A group of ten students can have cardiac contraction strength, heart rate and coronary blood flow demonstrated, as well as drug effects, using a single guinea-pig heart removed from the body. The reciprocity between agonists and antagonists can be visibly shown on isolated intestinal muscles. Thus, what is described in text books by mathematical formulae is demonstrated in living, graphic form. Anyone who has learned in practice how to introduce a cannula into a vessel in an anaesthetized rat, uses skill and care in doing this before he does the same thing to a human. The student learns from practical experience what lack of manual dexterity can lead to. This training in animal experimentation was routine for vascular surgeons where performing vascular sutures on rats maintained their manual dexterity.

Many practical experiments can be supplemented or replaced by filmed experiments or animation. The film cannot really offer direct experience, but it enables functions for which practical teaching would be too long and complicated to be illustrated. Anyone who has seen a film of the horror of morphine withdrawal in monkeys, will not easily forget the dangers of morphine addiction to humans. Similarly, anyone who has seen the calming effect of modern sedatives on nervous or aggressive wild animals confined in a cage, receives a living picture of the effect of these agents on the aggressive, mentally-ill patient. Modern pharmacological research work requires biochemical or biophysical methods; they are almost all too complicated or insufficiently graphic to be used as practical experiments. Thus, there is little need for animals in modern medical teaching.

Conversely however, it may be that students, who later wish to embark on research in medicine or biology, should also have the opportunity to do research on animals. Only thus do they learn *under guidance* how to handle experimental animals properly, methods of perfect anaesthesia and the most elegant way to perform an experiment. This instruction includes an appreciation of the importance of planning and preparation of experiments, their subsequent statistical assessment and interpretation of the results. Manual dexterity in taking blood samples, in deducing nerve impulses or in injection techniques require appropriate training. This cannot be learned from the television screen but only in practice. If this training is lacking during the instructional phase, then unsuitable experimental techniques will subsequently be used, they will be inadequately performed and animals will therefore be sacrificed needlessly.

The use of animals for teaching purposes has been markedly reduced in recent years. However, they should be retained where simple tests on small laboratory animals are valuable, particularly from the teaching point of view, and where the point cannot be better made by film or description. Furthermore, for the early training of all clinicians and biologists, who later enter a research field, there is a need for them to be used to handling experimental animals. It is very much to be hoped that this ability can be increasingly developed through special introductory and educational courses.

Reference
Stricker, S. Die experimentelle pathologie. Wein. klin. Wschr. 9 (1896) 959; 10 (1987) 425.

Sacrificed animals?
Animals in experimental surgery
U. Losert

The benefit of surgery following severe accidental injury, or in painful and common conditions such as gallstones, kidney stones, appendicitis, varicose veins and haemorrhoids is now taken for granted, and a reliable diagnosis, pain-free operation, uneventful postoperative recovery and long term success is expected. The media frequently carry stories about artificial hip replacements, heart surgery, cardiac pace-makers, artificial heart valves, vascular and brain surgery. We also hear about plastic surgery, which can restore disfigured faces, or reattach severed arms and legs in extremely lengthy and delicate operations. People are aware of the considerable progress in surgery in recent decades, but generally know nothing about the development of these methods. The public have certainly heard of animal experimentation, but do not realise that our current knowledge of physiological and pathological processes was mainly obtained from research with experimental animals. Today, this knowledge forms the basis for the recognition of the causes, diagnosis and successful management of functional disorders and diseases.

Animal experiments, particularly involving surgical intervention, are often regarded as vivisection. Thus even today, because of the dearth of information, many people have the wrong idea about the intentions, purpose, and performance of animal experiments. Until the middle of the 19th century, operations on man and animals really were vivisection, because the patient was generally fully conscious. Surgery was an emergency affair and usually performed by laymen. Thus, in 1837 at the Vienna General Hospital, Franz Schuh became the first doctor to be appointed head of a surgical department (Lesky, 1978). Even 50 years ago, patients were at risk of wound infection and blood poisoning since antibiotic therapy was unknown. Every major surgical operation, for example opening a body cavity, involved a risk. After many animal experiments, the introduction first of sulphonamides in 1936, and of penicillin in 1941, laid the foundation for modern chemotherapy. In general, it was not possible to perform lengthy operations in spite of the chloroform or ether anaesthetics already being used. The then scanty knowledge of pharmacokinetics and pharmacodynamics, as well as the side effects of the anaesthetic agents used, often led to untoward incidents. The then largely unmeasurable metabolic disorders intensified these dangers in very ill patients. Treatment of breathing difficulties using a combination of intubation and artificial respiration was first used in 1954/5 during the poliomyelitis epidemic. These advances in anaesthesia and internal medicine over the last 30 years were based on intensive research activity, including animal experiments, and today facilitate life-saving operations on the most severely ill patients of all ages.

Surgery as a profession, an art and a science (Wolner, 1982) can no longer be regarded in isolation, but is closely associated with the progress and the acquisition of knowledge in medicine as a whole, both in its principles and in the research employed. Surgery as vivisection, i.e. the cutting of living creatures, must consider the different functions of the single organ, its numerous interconnections and the superimposed regulatory systems of the organism such as the autonomic nervous and endocrine systems. Knowledge about complicated regulatory systems of the body facilitates both the surgical management and rehabilitation of the severely injured patient, and the early diagnosis and control of life-threatening, pathological bodily processes which progress almost automatically, e.g. the various reactions to shock. There are numerous examples of knowledge acquired through experimental animals which now contribute to the well-being of both man and animals. The researcher compounds and uses this knowledge, in anaesthesia and pain control for example, in his work with experimental animals. As early as 1894, the surgeon A. Fraenkl wrote: "Animal research is a proper and skilful surgical operation, which is carried out with every precaution and resource, just as is required of human surgery. First of all, anaesthesia in animal experiments is given the greatest possible use...."

The enormous technical progress of the last decades has occurred through the development of new measuring and diagnostic techniques (endoscopy, ultrasonic diagnosis, computed tomography, scintigraphy, nuclear magnetic resonance etc), as well as methods of medical laboratory investigation (immunofluorescence, hormone identification, metabolic investigations with isotopic markers etc). Insights into the body range for example, from the electron microscopic representation of a cellular component of the auricular muscle cells of the heart, to the chemical structure of the hormone-like active compounds they produce (atrio-natriuretic peptides). This blending of technology and chemistry in medicine has led to accusations of 'technological medicine' or 'poisoning by medical science': criticisms which are shown to be false and exaggerated by close consideration of the arguments, or examining the patient-doctor relationship.

Due to the introduction of technology, many questions in surgical research have answers which can only be resolved by animal experiments, thus requiring an increase in research activity. Nevertheless, the number of animals used in experimental surgery can clearly be reduced by refinements in measuring methods and data evaluation, establishing computer models, studies on perfused organs and cell cultures, optimal conditioning and preparation of experimental animals, and improving anaesthetic techniques and intensive care specific to the species of animal. With all these, the state-of-the-art of individual tests can be improved. Animal experiments in surgery should be regarded as investigations on the single animal. Experimental animals undergoing major and stressful operations need continuous observation and care, just like patients in an intensive care unit. Without intensive postoperative support of this kind, the experimental animal is not only endangered but the object and the skill of the test is also questionable.

The following overview of animal experiments in surgery, should help to evaluate their necessity or the use of possible alternatives, although it does not claim to be a complete picture.

Animal experiments for the training and education of surgeons
Animal experiments for surgical training are now rarely performed, and then only in exceptional cases. The acquisition of microsurgical techniques is associated with animal research, together with theoretical teaching and the use of cadavers. Here the surgeon must learn how to close the smallest vessels under the surgical microscope without loss of blood, as well as to aim for the most favourable and least intrusive form of anastomosis. With a vessel diameter of less than one millimetre, microturbulence and small surgically-occasioned abnormalities of the vessel wall may lead to occlusion, and can thus impede the successful reattachment of severed parts of the body. Furthermore, the introduction of a new and expensive surgical technique into clinical practice can necessitate a return to animal experimentation. Thus for example, heart/lung transplantation requires more specialised knowledge of thorax and heart vessel surgery, anaesthesia, diagnostics and therapy of rejection processes.

Many excellent surgeons have been trained without any experiments on animals. However, technically very difficult and complex surgical procedures, carried out by an inexperienced surgeon can endanger the patient's life. Nevertheless, this group of animal experiments should always be used to answer an open question also, e.g. the testing of new suture materials, determination of possible pharmacological side effects etc.

Animal experiments for material testing
These types of animal experiment are frequent in experimental surgery since materials which are introduced into an organism are recognised as being foreign. Body fluids and blood react very aggressively and lead to premature ageing processes through corrosion and erosion. Clot formation, fibrin and calcium deposits, connective tissue sheathing and inflammatory reactions can additionally contribute to a loss of function. One must consider the various requirements of surgical suture materials, e.g. the resistance to

infection, absorbability and material quality from high strength wire to microsurgery threads as fine as 0.002 mm. In addition, there are long term implants such as artificial hip joints, pacemakers with electrodes inside the heart, artificial heart valves and vascular prostheses. There are also the synthetic surfaces which come into contact with the blood in artificial kidney and heart-lung machines. Although specialist studies and long-term tests in different media - body fluids and cell cultures - can lead to elimination of certain substances in advance, the effects on the experimental animal can only be determined after lengthy studies. Negative or positive experiences acquired in clinical usage may also be considered as a matter of course. Many of these material tests could and should be combined with the formulation of further questions. There are no alternative methods which could predict the non-controllable occurrence of, for example, failure of an artificial heart valve or breakage of a vascular suture which might acutely endanger a patient's life.

Animal experiments to improve current forms of diagnosis and therapy
These animal experiments embrace a wide spectrum of experimental surgery. For example, radioimmunoscintigraphy has recently brought notable success in the non-invasive diagnosis of malignant tumours and their metastases. In principle, the method involves radioactively marking antibodies directed against tumour cells, injecting the radiolabelled antibodies into the patient and assessing the presence and extent of any malignant disease from the concentrated activity in the scintigraphic picture. The radioantibody technique is currently the focus of intensive research, and initial results in tumour-bearing mice are contributing to continuing improvements in the use of the method in humans. This investigative technique is designed for surgical application. Thus, by using the method in patients with breast cancer, all lymph nodes infiltrated by tumour cells may be identified and removed. Given greater experience and knowledge, the method will achieve widespread therapeutic use in the future, although together with cell culture research, additional experimentation in standardised animal models will also be necessary.

Flexible endoscopy equipment has now made it possible to examine internal regions of the body, and the possibility of local intervention has spared patients from stressful operations. The endoscopic removal of small gastrointestinal polyps can be used both curatively, for the purpose of early prevention, and diagnostically. The use of lasers in eye surgery has now become routine (e.g. retinal fixation for threatened detachment). Endoscopes with built-in lasers make it possible, following cadaver studies, for certain operations on the heart and its vessels to be performed in animal experiments without opening the chest. The physiological and morphological long term effects, as well as any possible damage caused by these operations, can only be assessed in living organisms. With the current state of knowledge their use in man is still too dangerous.

In restorative surgery, the need for muscle transplantation often arises. Various differing reactions and effects, such as vascular supply, connection and regrowth of nerves, baseline muscle tension and bleeding time play a major role in success. This knowledge is vitally important for clinical use and can only be acquired by long term studies in the experimental animal. This requires comparable conditions and constant periods of observation, as well as strength tests on prepared muscle with precisely administered direct nerve stimulation. These investigations, as well as histological and histochemical tests are not possible in the patient, since they would interfere with the successful outcome of the operation.

Almost 30 years ago, after intensive animal studies, the first pacemaker was given to a patient with cardiac rhythm disturbances. The electrode was implanted in the right heart chamber, and was connected by a long cable to the stimulation equipment which was standing on a table. After further active research, small, totally implantable devices were developed, which not only stimulated the heart muscle in disturbed rhythm, but were able to assess the individual's rhythm in healthy auricles. This has the advantage that the pacemaker can adapt itself to the heart frequency required. Previously, patients in whom cardiac impulse transmission was defective (sick sinus syndrome) could not be helped. In this situation, animal experiments helped detect other endogenous signals to regulate the pacemaker and thereby match the heart frequency to the needs of the body. Together with parameters such as acid-base metabolism, blood oxygen saturation, respiration frequency and the patient's bodily activity, the heart surgeon A. Laczkovics used the performance-dependent temperature profile of venous blood in the vicinity of the heart as an assessment parameter. After successful treadmill studies with dogs he submitted himself to two experiments of this type. Today, clinical investigations are carried out on volunteer patients.

Although self-experimentation is not infrequent in medicine - the Nobel Prize winner W. Forssmann demonstrated the diagnostic significance of the cardiac catheter on himself in 1929 - this method cannot always be recommended as an alternative. However, should the whole organism be necessary to resolve a problem, then man represents the only alternative to animal experiments.

Animal research in the development of treatment for previously untreatable diseases
In 1881, the great surgeon and researcher T. Billroth is said to have described the possibility of heart surgery as obscene. In the same year, the great clinician and scientist R. Virchow wrote "... the field of valvular heart disease, to which anti-vivisectionists refer solely and with not fully understandable aversion because of its incurability, is not completely barren..." On September 9th 1896, L. Rehn became the first surgeon to successfully suture a cardiac stab wound, and in 1925, H. Souttar successfully performed

a mitral valve repair. The significance of this advance was not appreciated at the time, and only in 1946, after improved understanding of haemodynamics and long term pathological effects was it reintroduced.

The development of cardiac surgery is so closely associated with animal research that many diagnostic methods and surgical procedures would be unthinkable without it. The decisive step occurred about 35 years ago when it became possible to disconnect the human heart from the systemic and pulmonary circulation and to operate both on the cardiovascular system and inside the heart itself. This advance was achieved with the help of two novel techniques; hypothermia and the heart-lung machine.

Actually, the English physiologist Starling had developed the 'heart-lung' principle in 1904, in which the experimental animal's heart, that of an anaesthetised dog, pumped blood through an artificial circulatory system. Thus, the ability of the cardiac output to adapt to requirements was recognised for the first time. Later, the mechanism of action of digitalis was studied using this technique. Although the method is no longer used, since for pharmacological studies, heart-lung models have been developed in rats or guinea-pigs, these methods formed the basis of many complicated, but clinically applicable, heart-lung machines.

The idea of J.H. Gibbon, which seemed utopian in 1934, of draining the heart of blood and substituting a machine for cardiac and pulmonary functions, led to the development of the heart-lung machine. Similar apparatus had already existed for the perfusion of isolated organs, which facilitated the study of metabolic processes. However, as well as draining all the blood flowing into the heart, the exchange of oxygen and carbon dioxide between the blood and respiratory air had to be achieved, since the maintenance of physiological values is vital for normal acid-base balance. The use of heparin for anticoagulating blood was known, but there was little detailed knowledge about the mechanism of blood coagulation and the effects of mechanical trauma on blood, especially in research animals. Thus, the heart-lung machine, in spite of generally successful animal experiments, could not be used clinically until the beginning of 1953, since severe postoperative blood coagulation disorders and frequent acute pulmonary failure were observed in dogs.

Hypothermia is a term used to describe a fall in body temperature, which in some animal species, e.g. the bat, leads to a physiological slowing of metabolism during hibernation. Animal experiments revealed that, with other mammals too, controlled cooling of body temperature from 28°C to 10°C is compatible with life, and leads to metabolic reduction with decreased oxygen consumption. In 1950, Bigelow reported successful animal experiments in which circulatory arrest for 6-8 minutes, resulting from a fall in body temperature to 28°C was also tolerated by the highly sensitive brain. Based on this knowledge, 1952 saw the successful correction of an inherited cardiac defect under hypothermia and circulatory arrest. Nowadays, both methods are used in combination; the blood in the heart-lung machine being cooled by a heat-exchanger and rewarmed towards the end of the operation. Subsequent extensive basic research into cardiac metabolism in cell cultures or in isolated organs led to the development of agents (cardioplegic agents, calcium antagonists etc.) which showed distinct cardio-protective effects during the blood depletion phase, and which are now widely used on humans.

This limited account should particularly clarify the need for animal experimentation, since without knowledge gained from such studies it would not be possible to deliberately induce cardiac arrest in a 'cooled' human patient, or to disconnect the heart and lungs from the circulation and divert the blood through a machine. No alternative is given here.

Animal experiments for the replacement of diseased or defective organs
It has been known for over 200 years that action potentials in nerves can be evoked and conducted by electrical currents and contractions produced on muscles. During technical development it became possible, like the cardiac pacemaker, to stimulate nerve functions directly. One method is to electrically transform and transmit impulses to the cerebral nerves and visual cortex of the brain using appropriate sensors, e.g. acoustic and visual impulses. Experimental animal research has since led to the first implantations in man which promise to be successful. Another method, developed in Vienna, is the mobilization of patients suffering from paraplegia by cyclic stimulation. In this method, several electrodes are affixed to muscle nerves and regulated by implanted microprocessors. Using this technique, functional breathing has already been achieved in five patients with respiratory paralysis by stimulation of the phrenic nerves. As a result of this procedure, continuous artificial ventilation in an intensive care unit was no longer necessary for survival. Movement processes and the restoration of the ability to move paralysed extremities present a considerably more complex set of problems. Here too, the first experiments on four patients with paraplegia have already been performed. As far as previous animal experiments and their clinical results are concerned, further research is required in both laboratory and animal experiments; for example in avoiding nerve damage, for determining the safe placement of electrodes to prevent mechanical and biological effects and elucidation of feedback mechanisms to improve the stimulation of several groups of muscles for controlled movements.

In surgery, the artificial replacement of bones, tendons, lenses, heart valves, and vessels have already become clinically routine. Nevertheless, further improvements are needed. Thus, obstructions often occur in lengthy vascular replacements particularly in the leg region, since the narrow blood flow and flexion in the joint area can, in spite of anticoagulants, lead to the formation of a blood clot. The latest laboratory investigations into certain blood cells and vessel wall lining cells (endothelial cells) have led to discoveries about their function, which also facilitate the cultivation of endogenous endothelial cells in blood vessels made of plastic. The ability of the endothelial cells, among other things, to prevent an accumulation of blood platelets on the vessel wall, could represent an important advance. With other organs, e.g. the organs of hearing and sight, muscles, liver, pancreas and heart, yet more research is needed, although clinical trials are already under way. The artificial heart, which has already proved satisfactory for short term use, is of particular interest. Just as the artificial kidney was used as a bridging aid until transplantation could be performed, the artificial heart has so far been implanted worldwide in 45 patients who were beyond recovery, and 22 out of 36 patients with transplants have subsequently survived.

The replacement of kidneys, heart, lungs, liver and pancreas by transplantation of donor organs is performed in specialist centres, in which the worldwide transplantation of over 150,000 kidneys and over 1,000 hearts has become routine. Transplantation surgery not only deals with the surgical and technical problems of organ implants, but also with the whole complex of physiological metabolic processes, organ conservation and immunological compatibility. Existing techniques involving cell culture and isolated organs may provide answers to many questions, but conclusions about physiological organ functions and long term effects associated with the body's regulatory systems are only possible on an intact organism. In this respect therefore, alternative methods developed for cell culture and perfusion of isolated organs do not represent replacements, but are an important part of the overall research. For specific problems therefore, the true alternative to the experimental animal is man.

Animal research to explain pathophysiological processes

Many pathophysiological processes are recorded in patients by careful clinical observations, although no explanation nor causal therapy is possible. For these problems, animal models in which the disease occurs naturally, or the appropriate changes can be controllably induced, may be used. The criticism is often expressed that an animal model is not transferable to man. This view can only logically be formed for some animal models after superficial consideration, if the attempt is made to transfer the result **directly** to man. Certainly, knowledge of the various physiological differences between the experimental animal and man makes it easier to compare and interpret their respective pathophysiological processes.

Large increases in blood pressure during pregnancy lead to acute danger to mother and child. Modern anti-hypertensive drugs are used to treat patients with high blood pressure, although their effects on the blood flow of the uterus and placenta are not clearly defined.

Investigations into mother and child cannot be performed on moral or ethical grounds. Although the placenta of a ruminant is very different from a human placenta in construction and shape (placenta syndesmochorialis multiplex), and gestation period (five months), the pregnant sheep has greatly contributed to our present knowledge of, for example, neonatal lung development (infant respiratory distress syndrome). On the basis of knowledge derived from animal species, it is possible to induce hypertension in the pregnant sheep and to examine the effect of antihypertensive agents on both mother and lamb. The diverse diaplacental transfer of drugs can now be investigated by perfusion studies on the isolated human placenta. These perfusion studies represent a way to reduce the number of experimental animals, but are not a true alternative for this type of problem.

Until recently, the spleen was routinely removed after traumatic injuries, until the high infection rate in these patients was eventually noted. There was also a increased tendency to thrombosis, and evidence of sepsis, especially in children. The changes in white blood cells and immunoglobulins seen after splenectomy were responsible, among other things, for the body's impaired defences. It is now known that the spleen fulfills important functions in blood cell maturation, lymphocyte formation, filtering particular elements and antigens from the blood and production of antibodies etc. Although numerous surgical methods and adhesion techniques (fibrin glue) are used to maintain the spleen, its removal in severe disorders is sometimes still inevitable. In man, various methods of autotransplanting functional splenic tissue have not had the desired result. The hope of these high-risk patients, who require constant antibiotic treatment, currently lies in systematic research in man and animal.

References

Bigelow, W.G., Callaghan, J.C. and Hopps, J.A. Amer. Surg. 132 (1950) 531.

Forssman, W. Klin. Wschr. 2 (1929) 2085.

Fraenkel, A. Über die Bedeutung des thierexperimentes für die chirurgie. Selbstverlag des vereines zur verbreitung naturwissenschaftlicher kenntnisse. 1894.

Gibbon, J.H. Minn. Med. 37 (1954) 171.

Laczkovics, A. Pace 7 (1984) 822.

Lesky, E. Die weiner medizinische schule im 19.Jahrhundert. Bohlau, 1978.

Rehn, L. Arch. klin. Chir. 55 (1897) 315.

Souttar, H.S. Brit. med. J. 2 (1925) 603.

Virchow, Über den werth des pathologischen experiments. Hirschwald, Berlin, 1899.

Wolner, E. Wein. klin. Wschr. 94 (1982) 140.

6

Provision for Research with Animals

Clear support is given here to animal research, because we are no longer dealing with captured mongrels or tortured monkeys. Nothing has contributed more to the correct maintenance and feeding of our domestic animals and livestock than information derived from modern research animals. Animal maintenance in satisfactory conditions is an incontestable prerequisite for the animal experimental researcher, just as sterile conditions in the operating theatre are for the surgeon.

Experiments on monkeys have mainly been transferred to specialist 'centres'. Experiments on dogs and cats have largely ceased because new methods permit the use of small laboratory animals, and thus specially bred strains of rats and mice. Of these animals, most are killed to obtain tissue for *in vitro* investigations. It is not the total number of animals used which gives a realistic picture, but the subdivision into species and a comparison with animals killed in other circumstances and for other purposes. From this it is evident how much the development of 'alternatives' is to the fore.

Is animal experimentation still necessary?
P. Skrabanek

Progress in life sciences would be unthinkable without experiments using live animals and animal tissues. Of 71 Nobel Prizes for Physiology and Medicine, 63 were awarded to scientists for discoveries based on animal experimentation (Ullrich and Creutzfeldt, 1985).

The rhetorical question in the title of this section can be answered 'no' only if the corollary question, "Is progress in life sciences still necessary?" is also answered with 'no'. This is the hub of the matter. Those who are opposed to progress in medicine and science are the lucky people who did not have a painful, incurable disease, or who do not have to nurse their own dying child.

Science, of which life sciences are a part, is not pursued with the sole aim of reducing human suffering; searching for knowledge is a human attribute, inseparable from man, just as speech, art or humour. Science is often criticised for being immoral. Henri Poincaré, whom Bertrand Russell called the most eminent scientist of his generation, wrote that science and ethics can never be in conflict because the domain of science (search for knowledge) and the domain of ethics (search for norms of conduct) only touch each other, but they do not overlap. In other words, science cannot be immoral, though some scientists pursue knowledge by immoral means, and, conversely, ethics cannot be scientific, although some moralists make absolute claims. Science chooses which goal to pursue, which horizon to push further, while ethics tell us by which means we are allowed to achieve it (Poincaré, 1904).

Ethical rules are meaningful only when the majority of people accept them as reasonable and agree to enforce them. It is silly to speak about 'animal rights' in a society which uses animals for food, clothing and sport. What is 'animal' anyway? Man is an animal. A fly is an animal. What rights should blue-bottles be allocated? Or should only cuddly mammals have rights? Should voles have the right not to be eaten by the fox, and mice the right not to be slowly killed by the cat? We live in an era of 'rights' at a time when the majority of mankind is denied the basic needs for a decent life. The emptiness of the language of rights was seen long ago by Jeremy Bentham who said that to speak about rights is nonsense, and to speak about natural rights is a nonsense upon stilts.

The increased public interest in the 'anti-vivisectionist' movement, together with the birth of a new branch of ethics which deals with the 'animal rights', is due to a variety of factors, some of them based on genuine fear and concern about the direction science is taking, while the others represent the dark, obscurantist streaks of anti-intellectualism opposing progress throughout human history. These two main strands, interwoven in the anti-vivisection propaganda, must first be disentangled.

Some experiments on animals are not necessary and are therefore indefensible. However, for a less informed member of the public, it may be difficult to decide what is necessary and what is not. For example, when the famous neuro-physiologist Brown-Séquard was given a sharp blow across the fingers with an umbrella by a lady who was present at one of his demonstrations, in which he was using a monkey, at the Collège de France in 1883, the lady did not know that Brown-Séquard's experiments enormously advanced our understanding of the function of the nervous system and the spinal cord. And how many more experiments will be necessary before we can offer hope to some paralysed victims (Anon, 1883).

Not all experiments on animals serve science, e.g. toxicity testing and safety control for substances used or consumed by humans. When such testing involves new cosmetics, lipsticks etc., it is not enough to have calls for the abolition of such tests from people who do not use such products themselves. Society must be informed about the nature of the tests and then decide whether they want more cosmetics or not.

Some animal experiments represent a useless repetition for the sake of producing 'research' papers which will be used by the author for filling the space under 'Publications' in an application for a better job. This can be prevented by establishing the competence of the researcher, by assessing the objectives of the proposed animal experiments, and by supervision. The mechanisms for this exist. For example, in England, inspectors from the Home Office, which is responsible for supervising adherence to the regulations about animal experimentation, made on average, 13 surprise visits a year to each registered centre for animal experimentation. This compares very favourably with the frequency of inspections of factories where considerable human hazard exists - about one a year (Paton, 1984).

While some animal experiments are unnecessary, the scientists would be the first to admit that much of the activity which passes under the name 'science' is not worthy of its name. Scientific literature is replete with peer criticism of slipshod research. Where animals (or humans) are used for unjustified, poorly planned experiments, such practices should be exposed, criticised and their repetition prevented.

A more complicated problem is the use of animals (or humans) for research in behavioural control. There is a potential gain in better understanding of the causes and treatment of mental diseases, although the use of animals as models of human mental disease appears to me absurd (Skrabanek, 1984). However, some of this research has been sponsored, directly or indirectly, by the military, and the implications are obvious: the results could be useful in controlling the minds of healthy people who are classified as 'enemies'. The public, and conscientious scientists, are rightly concerned about this type of research; and its nature and value should be scrutinised in informed debate.

Experiments which induce suffering and pain are also an area of contention. Proper use of anaesthesia answers some, but not all the objections. However, the anti-vivisectionists tend to forget that many diseases, whose causes and treatment the scientists try to discover by animal experimentation, also causes pain and suffering to human victims and their families. To avoid all suffering is impossible. The human lot is a tragic one, and will remain so, if the human race does not terminate the tragedy by annihilating itself.

While conceding all reasonable objections to unnecessary animal experimentation and to inflicting unjustifiable pain on animals, I would now like to turn to the second main trend in the anti-vivisectionist movement - the opposition to science; the opposition to the advancement of knowledge.

Since Adam was expelled from Paradise, we know that we cannot spit out the apple with the 'worm' of knowledge and revert to the state of blissful ignorance. Since Prometheus stole fire from gods, we know that we cannot put the gifts of Pandora back into her amphora. Rousseau's call to go back to nature, so dear to present-day utopians, appeals only to those who prefer wishful thinking to the harsh reality of the human predicament. The apostles of anti-science, such as Theodore Roszak, want us "to ground science in a sensibility drawing on the occult, mysticism, the Romantic movement..." (Wade, 1972).

Nature is mysterious and will remain so. For a scientist there is only one way: to stumble forward in the darkness without turning back. For anti-scientists, this would be a nightmare; they have a horror of not knowing, they have to deny ignorance by filling in the blank with wishful fantasy. As Erasmus observed, "man's mind is so formed that it is far more susceptible to falsehood than to truth....The fools are better off, first because their happiness costs them so little, in fact only a grain of persuasion, secondly because they share their enjoyment of it with the majority of men (Erasmus, 1512).

Acupuncturists do not have to do any animal experiments. They know it all. They understand the causes and treatments of all diseases. So do the practitioners of other 'alternative' medicines. But alternative medicine is no alternative. Roszak ridicules the objective knowledge of science: he and his ilk wish us to return to alchemy, astrology and irrational subjectivism. This is an ostrich-like attitude to human suffering. The 18th century English physician, Thomas Beddoes, put the following words under the heading "Experiments in medicine" in his notebook: "Those who decry them do not perhaps perceive that they cut off all hope from those at present incurable".

It is often alleged that animal experimentation in medical school is a deliberate attempt to desensitise future doctors. Surely they are more effectively desensitised by encountering their first 'patient' as a pickled cadaver. In fact, some desensitization is desirable: should a doctor faint when he sees his patient bleeding?

Compassion for humans is not necessarily accompanied by pity for humans. As pointed out by a *Lancet* correspondent, had the millions of human victims of Nazi-occupied Europe qualified under the laws protecting animals, as introduced by H. Goering, their fate might have been different (Seidelman, 1986). Some modern anti-vivisectionists would still prefer experiments on live humans than dead animals. This reminds me of a French surgeon who taught Western medicine in China and asked for some corpses which he could use for dissection. This request was received with horror by his Chinese employees, who nevertheless, assured the surgeon that he could have an unlimited supply of live criminals (Russell, 1950).

Recently, the Greater London Council allowed the British Union for the Abolition of Vivisection and the National Anti-Vivisection Society to erect a statue in a public park to a dog, which in 1903 (according to the inscription) "endured vivisections....till death came to his release" (Anon, 1985). The inscription did not mention that the dog was operated upon (always with anaesthesia) by E.H. Starling, W.M. Bayliss and Henry Dale, most brilliant British scientists, and that the experiments led to the discovery of the first hormone and to the birth of modern endocrinology.

The most powerful argument of the anti-vivisectionist extremists, who campaign for abolition of all animal experiments without exception, is presented in books by Hans Ruesch. I own the American version (Ruesch, 1983). It is a *j'accuse* type of book, in which no hold is barred if it serves the Cause. It is a book which makes converts readily, including medical doctors. (This is not surprising since doctors have to endure dogmatic eduction which discourages critical thinking.)

Much of what Mr. Ruesch says is true, but it is half-truths which he exploits with great effect. He accuses scientists of "greed, cruelty, ambition, incompetence, vanity, callousness, stupidity, sadism and insanity". I have seen it myself, but which category of people is immune to these charges, including writers to which Mr. Ruesch belongs? Mr. Ruesch finds it hilarious that studies of the "love life of the flea" and of "the mating call of the mosquito" attract funding. Is he not aware that the fleas are the vectors of the plague, and the mosquitos of malaria? Better understanding of their reproduction could save millions of human lives. Some of Mr. Ruesch's accusations are malicious: "insulin treatment has done more damage than it brought benefits, has killed more people than it has saved". He **has to** say this, since insulin was discovered by animal experimentation, therefore such a discovery must be a Pyrrhic victory for scientists. Mr. Reusch's alternatives are bizarre: "medical science today knows nothing with certainty that Hippocrates didn't know already". He does not say that Hippocrates humours (blood, phlegm, yellow bile and black bile) have now only humorous value. Not surprisingly, Mr. Ruesch approves and recommends homeopathy, osteopathy and acupuncture: "they raise medical art gradually up to the Hippocratic level again".

Fortunately, not all animal welfare groups hold such extreme views, and most of them have an important role to play in finding the proper balance between animal welfare and human needs. In Great Britain, the Cruelty to Animals Act of 1876 has become inadequate for regulating animal experimentation, and a new version, known as the Animals (Scientific Experimentation) Bill is now being debated in the House of Commons. The British Veterinary Association, the Committee for the Reform of Animal Experimentation, and the Fund for the Replacement of Animals in Medical Experiments have welcomed the Bill. As these organizations jointly stated, what we need is an effective compromise between the welfare of animals, the legitimate demands of the public for accountability and the equally legitimate requirements of medicine, science and commerce. In a reasonable society a reasonable compromise must be found.

References

Anon (1883). J Am Med Assoc 1, 28.

Anon (1986). A new anti-vivisectionist libellous statue at Battersea. Br Med J 292, 683.

Erasmus of Rotterdam (1512). In praise of folly. Translated by B. Radice. Penguin Books, Harmondsworth, 1971.

Paton W. (1984) Man and Mouse. Oxford University Press, Oxford.

Poincaré H. (1904). La Valeur de la Science. Flammarion, Paris.

Ruesch H. (1983). Slaughter of the Innocent. Civitas Publications, New York.

Russell B. (1950). An outline of intellectual rubbish. In: Unpopular Essays. Allen and Unwin, London, 1950.

Seidelman W. E. (1986). Animal experiments in Nazi Germany. Lancet i, 1214.

Stock J. E. (1811). Memoirs of the life of Thomas Beddoes. John Murray, London.

Skrabenek P. (1984). Biochemistry of schizophrenia: a pseudo-scientific model. Integrative Psychiatry (New York) 2, 224.

Ullrich K. J. and Creutzfeldt O. D. (1985) Gesundheit und Tierschutz. ECON Verlag, Düsseldorf and Wein.

Wade N. (1972). Theodore Roszak: visionary critic of science. Science 178, 960.

The question of responsibility
Ethical aspects and legal requirements
N. Zacherl

The ethics and provisions of positive justice regulate the circumstances of life in which people participate, either in the form of action or failure to act. This action, or lack of it, is evaluated and - at least in positive justice - may result in the imposition of sanctions.

The present focus of interest is animal research. Although it may seem strange, the precise definition of an animal experiment, at least in the legal sense, remains unclear. For example, the 1974 Austrian law on experimental animals evades responsibility for this question in only legislating for certain kinds of experiment, namely those associated with pain or suffering for the animal and performed for certain purposes. Nevertheless, legislation seems to recognise other animal experiments since it would otherwise have been meaningless that the draft of a new Viennese regional animal protection law (June 1986) prohibited animal experiments "inasmuch as they are not expressly permitted by other legal provisions", where comments pertaining thereto refer to the 1974 law mentioned above. The problems of animal research which arise from a segregation of the laws on animal experiments in particular, and animal protection in general, further aggravated in Austria by these being the responsibility of constitutionally diverse legislative bodies (state and regional), will be mentioned, but not discussed in detail.

What then constitutes an animal experiment from an ethical or legal standpoint? An animal experiment is basically only one of several ways in which animals are employed. Thus, an animal may be used - sometimes after appropriate training - to perform a specialised task (e.g. packhorse, hunting dog, police dog, guard dog etc.), for sporting purposes, for some physiological requirements (e.g. to provide food, clothing, medicines such as sera or particular organs which are used in medicines), to provide for emotional needs such as affection, company (domestic pets) and so on. The killing of animals which man considers undesirable (predators, pests) is also a form of man-animal interaction which should be mentioned here.

The important aspect of the animal experiment as a way of using animals is man's use of them to increase his knowledge of biology or to test scientific hypotheses which are not themselves in the experimental animal's interests. Thus, an animal is used in a manner unacceptable to it, without the intended outcome being guaranteed or necessarily profitable. Kant spoke in *Metaphysic of Morals* #17, of the "excrutiating physical experiments performed in the name of speculation".

As a philosophical discipline the ethic bases its finding upon intellect, and its goal is to achieve actions based upon reason. In this way the ethic traditionally claims to give legitimate grounds for correct decisions, and thus to justify the attempt of ascribing general validity to that which one has recognised as right. It expresses the obligation for the realisation of true humanity, and touches not only on the question of good will but also of the right and fair way.

Today almost nobody denies man's responsibility for his environment and thus also for animals. However, on what grounds of reason (and not of religion) is this human responsibility based, which takes responsibility for the protection of the individual animal but not the species? The anthropocentric answer to this question arises from a dualism which considers humans as the subject and animals as the object, which consequently have no individual personal rights. Thus, animal protection is only permissible insofar as it serves mankind; the animal is exclusively a medium for achieving man's goals. So Kant (*Metaphysic of Morals* #17) established the obligation to abstain from violent and cruel treatment of animals, since compassion for the suffering of animals would otherwise be blunted, thereby "undermining a natural tendency to morality in behaviour towards other people". In this category belong all other reasons for animal protection which require the maintenance of animals' wellbeing in man's interests e.g. in order to obtain the maximum information from an animal or to protect human life. Thus, the veto on animal torture ultimately rests in man's duty to himself.

A more plausible, non-anthropocentric tenet of animal protection, which relates to the animal itself, stems from animals' capacity to endure suffering. It is based on the principle, founded on reason, which esteems those interests, which man considers legitimate for himself, worthy of protection in all other creatures who are in the same situation. Pain and suffering for the animal seem to be an evil which have to be avoided out of respect for the animal's basic rights. Man's responsibility for an animal therefore increases in accordance with its capacity for suffering. Thus, whether or not an animal has any rights in this connection is irrelevant. Man's duty to ensure that an animal is free of pain comes into insoluble conflict with his own interests, if - as in animal research - there is no other way of conducting an experiment other than subjecting an animal to pain or suffering. In this case a decision must consider whether the ontological superiority of man and the qualitative difference - arising from a difference in consciousness - between the relative capacities of man and animals for pain and suffering can justify the inequitable treatment of the animal.

Basically, in animal research there is a conflict between the stress inflicted and the useful knowledge which may be gained. The decision on the admissibility of a particular animal experiment is based on the supposition that certain animal experiments are necessary and permissible in principle. The acceptability of an animal experiment which is associated with a degree of pain and suffering, and to some extent the probability of its benefit to man, can only be substantiated by moral/ethical considerations of the results in individual cases. In addition, because of the animals' *prima facie* claim that their existence should be as pain and suffering-free as possible, individual evaluations and decisions should always be made (Patzig, 1986). The burden of proof in these cases always rests upon those seeking justification for performing animal experiments, since it is their aim to relegate the animals' *prima facie* case to subordinate status. The following arguments should help to assess the potential value of an animal experiment:

- that only analogous benefits may be balanced against each other,
- that the hypothesis to be tested in the animal experiment is important, and that there is a possibility of its confirmation or refutation,
- that there is a need for the animal experiment (i.e. that there is no other way of confirming or refuting the hypothesis),
- that it will be possible to transfer the results of the animal experiment to the human situation,
- that a method appropriate to the hypothesis is employed,
- that the stress inflicted on the animal by the experiment is reasonably proportional to the importance of the hypothesis,
- that the number of animals used in the experiment, and the stress inflicted, is minimized.

If the degree of pain and suffering inflicted in an experiment is one of the accepted criteria for judging the admissibility of an animal experiment, there are still two aspects to be considered:

- The fact that something is perceived as unacceptable or even repugnant, is not itself a basis for regarding it as wrong.
- The accurate evaluation of pain and suffering in animals presupposes the recognition and appraisal of a wide range of parameters by an educated and experienced observer.

Discussion about the admissibility of animal research should not affect the view that, especially in medicine, performing an animal experiment can be ethically justified, perhaps as a prerequisite in the diagnosis of a particular patient. Indeed, one may even go so far as to suggest that the researcher, who has the necessary facilities, also has a duty to perform essential animal experiments, perhaps to develop treatments for incurable diseases. A doctor for example, would thus contravene his professional obligations if he failed to use animal research for diagnostic or therapeutic measures, or the determination of toxic effects on man and animals. This is also true of fundamental medical research, since he must use all available research tools to alleviate human suffering, to cure diseases and to prevent damage or injury to his patients, and thus in the wider sense, to mankind (e.g. infection prophylaxis and environmental damage).

Although almost all regional legislation on animal protection, and the need, admissibility and manner of performing animal experiments exists expressly to protect experimental animals, the precise legislative details and their provisions for conducting animal experiments are different in individual countries. Reference can be made here to the very clear presentation of differing national regulations produced by the US Congress Office of Technology Assessment in February 1986. Obviously the differences between the national regulations also determine the initial situation in research and development in individual countries. It would be distinctly hypocritical if animal research were prohibited in one country, but if that country nevertheless accepted the results and benefits which had been developed with the help of animal experiments. Thus, very restrictive animal research legislation, taken seriously, must finally lead to renunciation of the use of specific, animal research-based knowledge and results.

One benefit of European consultation is that a step has been taken towards standardisation in animal research legislation (draft of 18.3.1986), by establishing a convention for the protection of vertebrates for research or other scientific purposes. Within the European Community, the member states have committed themselves until November 1989 to standardise their national regulations according to agreed principles.

In Austria, a special Federal ruling was made on animal research in 1974 although detailed animal protection regulations have existed previously on a regional level. This law makes it necessary to hold a licence to perform experiments involving live animals which are associated with pain or suffering for the animal and which serve the purposes of science and development, scientific education, medical diagnosis or the testing and proving of drugs, foodstuffs, insecticides and herbicides, cosmetics etc. Unlike the 1977 Danish animal research act, research which only causes slight transient pain is not excluded from licence control according to the European Assembly Convention and European Community guidelines. The Austrian law also contains the provision that an animal experiment may only be authorised and carried out if a 'legitimate interest' in it can be authenticated. Thus, the Austrian ruling goes far beyond other national or supranational regulations (in Switzerland for example the justification for a research project does not generally conform to the European Assembly Convention, and European Community guidelines are only applied where very painful experiments are involved). As with the animal protection law of the Federal Republic of Germany, the only animal research not requiring authorisation is that which is undertaken in certain state research establishments, within the legally transferable sanctions of those establishments, or on the basis of legal or judicial directives or of agreed testing of sera and vaccines. In order to prevent a conflict between a legally binding regulation for an animal experiment, and the possible rejection of a relevent sanction, these exceptions seem sensible. Nevertheless, it would be worth considering a guarantee that these non-licensed animal experiments would only be carried out in accordance with the legally prescribed and established requirements for personnel and equipment.

Killing an animal for research purposes (e.g. for the removal of organs) does not constitute notifiable animal research in Austria, since the 1974 animal protection law - like the 1978 Swiss act - speaks of 'living animals'. The English animal protection law - Animals (Scientific Procedures) Act 1986 - and the European Assembly Convention state explicitly that killing for research purposes does not represent notifiable animal research.

The view that every individual animal experiment requires a special licence is not embodied in the 1974 Austrian animal protection law in spite of official practice to the contrary. There is continual discussion about the sanctioning of animal experiments. Furthermore, the provision that a licence must determine the type of animal experiment and the person(s) conducting them, points clearly towards a licence possibly embracing a series of experiments or entire projects. The Federal German ('projected research'), the English ('a programme of work') and the Swiss regulation (animal protection order 1981: 'type of research or series of experiments') are certainly more precise.

In Austria, permission to perform animal experiments is conferred on individuals or corporations; in the Federal Republic on an individual (person) or institution; in Switzerland only on directors of institutions or laboratories. The English law recognises a 'personal licence' and a 'project licence', which are both only permitted to physical persons. Thus the 'personal licence' qualifies its holder to undertake specific research on specific animals in specific institutions. The project licence on the other hand, primarily gives authorisation to carry out a particular programme of research, and is granted to someone who undertakes full responsibility for this programme. This dual system has the advantage that prescribed individual provisions do not have to be checked for every licence. Moreover the English law requires that a project licence is only granted to an establishment which is recognised as a 'scientific procedure establishment'. Thus, the criteria regarding the institutional facilities and the lawful care and contol of the experimental animal need not be re-examined for each individual licence application.

According to the 1974 Austrian animal experiment law, animal experiments may only be performed to prevent, understand or cure diseases in man and animals, to obtain scientific knowledge and for purposes of scientific training. Compared with other legal requirements these justified aims of animal research are imprecise, and also narrowly conceived. It is noticeable that the European Assembly Convention and the English law on animal research also allow for botanical maintenance and the aims of environmental protection (the latter is also included in the German animal protection law). The weighing of the anticipated pain and suffering against the potentially beneficial outcome of an animal experiment are expressly mentioned in German and English law as prerequisites for permission; in Austria the allusion in the animal protection law to the 'legitimate interest of the experiments' can be understood in this sense. The Swiss law moreover makes the sanctioning of an animal experiment dependent on the suitability of its methodology for achieving the object of the experiment. International regulations usually include the provisos that animal experimentation may only be performed if the object of the experiment cannot be achieved by any other methods or procedures, that animal experiments may only be undertaken by specially qualified persons, and that all unnecessary pain and suffering associated with the experiment must be avoided. As to the obligation to keep records and results, there are provisions which require relevant internal notes to be kept which can be inspected on behalf of the authorities (Austrian, German, Swiss) or of such persons who demand public accounting (England, European Assembly Convention). The publicly accessible information concerns the actual species and number of experimental animals used and the type of research carried out; disclosure of the results of an animal experiment is provided for neither in the legal system discussed above nor in the European Assembly Convention. The latter contains however the signficant reference that the treaty countries should as far as possible recognise the results of animal experimentation from other countries.

To summarise, it has been shown that in positive law, the ethical problems of animal experimentation, which arise from the expected benefits relative to the pain and suffering of the experimental animal are regulated with similar intent, albeit in different ways, so that the legal standardisation adopted by the European Assembly Convention and the European Community seems fair and expedient. Thus, the English system of a three-part authorisation (approval for the facilities for animal experimentation, private licence and project licence), which has already found its way into European Community guidelines, is worth considering.

References

Auer, A. Umweltethik - ein theologischer beitrag zur ökologischen diskussion, 2. Aufl. Patmos, 1985.

Bochenski, J. Zur ethik der tierversuche. Sandoz-Bulletin Nr.75, Oktober 1985 Sandoz AG, Basel.

Böckle, F. Das tier als gabe und aufgabe. In Händel, U.M.: Tierschutz, testfall unserer menschlichkeit, Fischer, 1984.

Böckle, F. Grundbegriffe der Moral, 8. Auf. Pattloch, 1977.

Erbrich, P. Auf der suche nach einer umweltethik, Orientierung, Nr. 6/49. (Institut für Weltanschauliche Fragen, 1985) Seiten 68ff.

Guidelines for the recognition and assessment of pain in animals. The Veterinary Record, 118, 12 (1986) 334.

Höffe, O. Ethische genzen der tierversuche. In Händel, U.M.: Tierschutz, testfall unserer menschlichkeit, Fischer, 1984.

Patzik, G. Ethische aspekte von tierversuchen. In: Vorträge zum thema mensch und tier. Studium generale/tierärzliche hochschule Hannover, Band 3/4 (M. u. H. Scaper, 1986).

Reiter-Thiel, S. Die psychotherapie und die ethik. Falsche oder notwendige verbindung? Die Presse vomm 22/23, November 1986, Spectrum, Seite X.

Schoch, M. Die Mitkreatur, N.Z.Z. vom 12 April 1985, Fernausgabe Nr.83, Seite 25.

Sprigee, T.L.S. Philosophers and anti-vivisectionism. ATLA 13 (1985) 99.

U.S. Congress, Office of Technology Assessment, Alternatives to animal use in research, testing and education, insibes, Kapital 4 (ethical considerations) und 16 (Regulation of animal use in selected foreign countries) und anhang E (International agreements governing animal use). U.S. Government Printing Office, 1986.

Weibel, E.R. Tierversuche: Der ethische konflikt des wissenschafters, N.Z.Z. vom 23 November 1985, Fernausgabe Nr. 272, Seite 28.

Suitable or not?
Modern maintenance of research animals
H. Juan

This type of article would be one-sided if it were to exclusively deal with the maintenance and care of research or laboratory animals (Figure 19). For this reason the maintenance of domestic animals and livestock are briefly considered.

The main aims of the article are to briefly present the present state of animal maintenance and care as well as their 'use', and to discuss the various different requirements. Compromises certainly have to be made. Our ethical responsibility for the animal involved *must*, however, always be concerned with the maintenance of animals as *conditio sine qua non*.

Demands from the animal protectionist' point of view
The demands of animal protectionists range from limitation, through complete abolition to the fundamental rejection of the killing of an animal. The latter point is a tenet of belief, but is not rooted in the Christian ethic. Everyone has the right to make these

demands in principle, but not to force their attitude on others, particularly without proven justification.

He tempts the pooch with a pretzel
The young lady calls to it to put it down

Figure 19. Even today, some people imagine the procuring of experimental animals to be as described in this sketch by Wilhelm Busch. However, the institute's laboratory assistant is not a dog-snatcher and the fat dog would, in any case, be of little use for any experiment. Those complicated studies for which there is no alternative to dogs are now only performed in specially reared breeds of laboratory dog.

One may understand the role the dog snatcher once played when one sees the countless dogs straying around on the outskirts of a Brazilian city. Undernourished, scavenging, swarming with vermin and parasites, mostly sick, they represent a primary hygiene problem. Whose worry is that?

Has man the right to keep animals and 'use' them in any way he likes? The answer must indeed be a clear *yes*. Since prehistoric times people have lived directly or indirectly on animals, and the evolution of mankind would otherwise not have been possible. More than 12,000 years ago the domestication of some species of animals began, firstly with dogs and goats. In time, all the types of domestic animals followed. More recently they served to ease, maintain or improve living conditions, for example, for pleasure, 'sport' and so on. To maintain the lives of humans, animals must generally give up their own. Without animal food products the survival of man, and of many domestic animals, would not be assured. If this point is accepted, it must also be permissible for animals to be killed for the improvement or safety of human health. Nevertheless, the demand that every animal death - be it in the slaughter house or the laboratory - should be carried out as humanely as possible is absolutely justified. Equally justified is the attempt to replace animal experiments, which are associated with unacceptable suffering, with better methods.

There are numerous animal experiments which cannot presently be replaced (Zell, 1984). This is because of legal requirements and their contribution to drug safety, important pure research and surgical methodology. All such experiments are *always* carried out painlessly under anaesthesia. In many cases, animals are needed (for organs, cells, blood etc.) even for important 'alternative methods' (for definition see Zbinden, 1985). Thus, it is to be concluded that special accommodation and space for experimental animals should be available in institutes in order to be able to guarantee suitable breeding, maintenance and care.

Requirements from the animal researchers' point of view
The most important requirements (this is not an exhaustive list) include the following points :

1. Healthy animals
2. Specific species of animal for specific tasks
3. Hygienically and genetically defined breeds
4. Good facilities for breeding, maintenance and care before and after the experiment.

Point 1. Animals *must*, even for organ removal, be physically and mentally sound. Only then will experiments be reproducible and the resultant data useable. Thus, considerably fewer animals will also be needed.

Point 2. In spite of the essentially similar 'ground plan' there are many well known differences which must be considered, e.g. for extrapolation of the results to humans (see p. 31). Although some undesirable side effects of drugs in animal experimentation are never encountered - especially those which have nothing to do with the 'main effect' - studies nevertheless show an agreement of all adverse effects between animal and man of 70%-80% (see *Bundesverband der Pharmazeuticschen Chemie*, 1985).

Point 3. Because of their homogeneity, hygienically (e.g. specifically pathogen-free animals) or genetically (e.g. in-bred) defined animals certainly reduce the number of animals needed for reliable results, or are otherwise required for special projects (e.g. in immunology).

Point 4. This requirement is a basic prerequisite for points 1-3 as well as for obtaining organs for many 'alternative methods'. In addition, the post-operative care of animals - similar to that required for humans - can only be performed in appropriate experimental animal establishments.

Utilisation and 'expenditure' of laboratory animals
Apart from killing for human and animal food purposes (e.g. for dogs, cats and zoo animals), animals in refuges are put to sleep (euthanasia), die on the streets, or are 'destroyed', sometimes very painfully, as 'pests' (Table 4). Laboratory animals are also killed in large numbers (Table 5). These tables are not intended to trivialise the high 'expenditure' of laboratory animals in the pharmaceutical industry or in research, but rather to demonstrate that the number of animals killed directly for human food is considerably higher. In addition, large amounts of meat are needed for the benefit of domestic pets (cats and dogs). In Vienna, it has been calculated that over 20,000 tons of meat are needed annually for its 70,000 dogs and 100,000 cats, This corresponds to almost 50,000 cattle or 100 million rats, Furthermore, wild animals apart, considerably more tame animals are shot by hunters than are used in research.

Many laboratory animals, which are bred and kept especially for pure research, are not research animals in the narrower sense. They should be regarded rather as animals for slaughter, since they are killed painlessly. The aim of these procedures is to obtain organs, tissues or cells. Thus, 'research animals' in the narrow sense includes animals which are specifically used for *in vivo* research. However, there is no difference in breeding and maintenance for either group. For this reason, modern 'research animal establishments' should not be rejected by animal protectionists, since they are a main source of the so-called 'alternative experiments'.

Requirements of animal maintenance
Obviously, species-adapted maintenance is the only proper way! However, what is meant by species-adapted? Nobody can define this completely for every species of animal. Wild animals can be observed and their modes of behaviour and habitat studied. This leads to an accumulation of knowledge, which cannot easily be transferred to domesticated, human-dependent animals. Even wild animals vary their habitats and adapt themselves to human civilisation (e.g. birds, coyotes, raccoons and rodents). Thus, it can be concluded that animals are capable of adapting and that the concept 'species-adapted' is too broad.

Table 4. Numbers of animals killed for different reasons (examples)

Cause of death	Number of animals p.a.		Sources
Slaughtered 1983/84	44,211,100	cattle calves pigs sheep	BML, Statist Annual Report 1/85, in Pro & Contra Tierversuche, Bundesverband der Pharmazeut. industrie, Fed.Rep.Ger.
1985	150,000	(see above)	Vienna
Road casualties	36,621 27,319 12,807 1,308	hares deer pheasants partridges	Hunt stats.1984/85 Austria (OAMTC-journal Oct 1985)
Shot by hunters	100,000	cats (c.)	Official Hunt Stats for Schleswig-Holstein and Lower Saxony
	> 250,000 > 25,000	cats dogs	Ullrich and Creutzfeld 1985, according to Deutschem Tierschutbund 1985, Fed.Rep.Ger
Agricultural activities	100,000 60,000 200,000 50,000	hares deer pheasants partridges	Assessments of Schutzgemeinschaft Deutsches Wild, in Ullrich and Creutzfeld, 1985, FRG
In animal sanctuaries	> 100,000 > 100,000	dogs cats	Ullrich and Creutzfeld (1985) according to Mitteilung der Bundesregierung FRG

The Table is rather incomplete and should only be taken as examples. For example, shooting of game or destruction of 'pests' are not included. Although comparisons are certainly emotive, it must be said - backed up by the USA example - that for every cat killed in a laboratory experiment, 50-60 domestic animals (dogs and cats) are put to sleep by euthanasia. Another calculation reveals that every year, per head of population, 0.2 research animals are killed, but 10 animals are slaughtered for food (Profile no.47, 18.11.85).

Table 5. Research animals, laboratory animals. 40 drug companies are considered in this statistic. The numbers include animals used for drug research, development, quality control and production in the Federal Republic of Germany, and include contract research as well as those animals which were killed painlessly to obtain organs and cells. The number of animals used for other research as well as other purposes (production of sera or vaccines) is equal to, or less than, the numbers quoted above.

'Research animals' used in 1984 in the German Federal Republic according to the Bundesverband der Pharmazeutischen Industrie

Mice	1,371,707
Rats	824,560
Guinea pigs	80,784
Rabbits	56,964
Hamsters	8,375
Dogs	8,191 *
Cats	5,255
Pigs	1,443
Cattle	724
Monkeys	592
Horses	2
Other animals such as sheep, goats, newly hatched chicks, fish, frogs	86,164 **

* Dogs are particularly used in circulation (e.g. blood pressure) research.
** In 1979 > 900 sheep and >400 goats were used.

Similarly, research projects on the maintenance of domestic animals have led to discoveries which has sometimes resulted in an improvement of animal maintenance (see pp. 184-194).

There is some legislation on animal maintenance within the framework of animal protection. In Switzerland, for example, they embrace all species of animals, including laboratory animals.

Useful, unbureaucratic guidelines, even though not written up in detail, for the sale, transport, accommodation and feeding of research animals were given by the Council for International Organizations of Medical Sciences (CIOMS).

Adequate guidelines, which should be binding for research animal maintenance on large or small scales, were established by the Gesellschaft fur Versuchstierkunde (GV-Solas) (Society for the Investigation of Research Animals). They comprise, based on observations and years of experience, recommendations which *should* represent the best in all areas. Improvements are currently taking place.

Quality of (research) animal maintenance
Even in the best animal maintenance there is a conflict between 'should be' and 'is'. The 'should be' circumstances would be species-adapted maintenance, the 'is' circumstances are often miserable - due to financial strictures or purely from ignorance. In agriculture, although much improved, some problems still remain. It is sometimes particularly bad in the private sector, when animals are kept in miserable conditions out of ill-understood 'love', and fed to death. The maintenance of research and laboratory animals on the other hand is often better than that of livestock and domestic animals. It is also understandable, even with the best of intentions, that it is not possible to attain the absolute best. There is always a certain economic limit.

The recommendations of GV-Solas for all areas of animal maintenance can however be considered quite exemplary. They embrace the following areas: organised forms of research animal maintenance such as centralised or decentralised establishments and research animal farms, space requirements (open, closed and isolator systems), personnel requirements, techniques (air-conditioning, lighting, building materials etc), costs, specific animals for specific purposes, breeding, hygiene and feeding.

Detailed accounts of individual points or of special kinds of maintenance such as quarantine quarters, closed systems with reliable environmental hygiene barriers (maintenance of specifically pathogen-free animals) will not be given here. Modern research animal establishments understandably contain various systems of this kind. Instead, the interested reader is referred to relevant literature such as that quoted in the 'Recommendations for planning, structure and erection of research animal areas by institutions carrying out animal experiments' (GV-Solas).

Good research animal establishments work according to the recommendations of GV-Solas. The correct guidelines for space and surface sizes are, for example, given so that they *can* be decreased or increased. By observing these assessments no disturbance to the animals' wellbeing is seen. Some values such as those for group maintenance of dogs in boxes, are given with the proviso that an adaptation is warranted by appropriate rearing; some (enclosed holdings) can be reduced under the conditions described above.

The feeding of animals should also be as good as possible. To this end, a great variety of standardised food mixes have recently been developed, whose composition complies with every nutritional requirement. In most cases we are talking about dried food pellets which can be stored much better and in less space than any other raw fodder. Although it does not correspond to natural feeding, as far as can be ascertained, it does contain all essential food components, including minerals and vitamins.

Again, good maintenance certainly reduces the number of research animals, since only absolutely healthy animals (physically *and* mentally) may be used to achieve worthwhile results. From this point of view, the maintenance of research animals can be considered suitable and sensible. It is however certainly not perfect, as no animal maintenance can ever be. Further improvements are needed and will be undertaken continually with increasing knowledge. The usual maintenance of livestock and tame animals can also profit from these experiences.

References

Auto Touring (ÖAMTC): Wildtöter, Okt, Nr. (1985).

Bundersverband der Pharmazeutischen Industrie (BPI). Tiere in der Arzneimittelforschung, Eigenverlag (1982).

Bundersverband der Pharmazeutischen Industrie (BPI). Pro & contra tiervesuche. Arbeitsgruppe tierschutz. Eigenverlag (1982).

Bundersverband der Pharmazeutischen Industrie (BPI). Leben mit dem neuen tierschutzgesetz. Symposium in Bonn/BRD, vom 15, 5. 1986 Eigenverlag.

Council for International Organisation of Medical Sciences (CIOMS). Internationale leitprinzipien für die biomedizinische forschung mit tieren. Eigenverlag.

Gärtner, K. Auf tierversuche kann noch nicht verzichtet werden. Umschau 7 (1983) 213.

Gergly, S.M. and Michalec, H. Wer folter duldet, foltert mit. Profil. 47 (1985) 60.

Gesellschaft für Versuchstierkunde (GV-Solas). "Empfehlingen" Bd. 1-9, Eigenverlag (1977-1985).

Höfer-Bosse, T. and Scharmann, T.W. Numbers of animals used in toxicological experiments - with particular reference to the Federal Republic of Germany. ATLA. 13 (3) (1986) 217.

Schweizer Bundesrat. Botschaft über die Volksinitiative "für die abschaffung der vivisektion" (30. 5. 1984).

Sechzer, J.A. The role of animals in biomedical research. Ann. N.Y. Acad. Sci 406 (1983) 1.

Smyth, D.H. Alternativen zu tierversuchen (Hsrg. Spiegel, A.) G. Fischer Verlag, Stuttgart, 1982.

Tierschutzgestze. Tierversuchsgesetz, tierschutzverordnungen, erläuterungen, novellen, gesetzesentwürfe von: Bundesrepublik Deutschland, Dänemark, Österreich, Schweiz (Stand: September 1986).

Ullrich, K.J. and Creutzfeldt, O.D. Gesundheit und tierschutz. Econ. Düsseldorf 1985.

Zbinden, G. Alternativen zum tierversuch. Neue Züricher Zeitung 174 (1985) 29.

Zell, R.A. Der mensch. Herr über Leben und Leiden. Bild. d. Wissensch. 12 (1984) Sonderdruck.

Demand for existing alternative methods
Promotion of new alternative methods
F. Lembeck

Man and animals
Attitudes to animals have, at different times and cultures, been very diverse. The cow is sacred in India, while in all other parts of the world it is livestock. The Moslem must cleanse himself after contact with a dog, while we regard the dog as a domestic pet and as a hunting animal. The dove is the symbol of peace, yet is a nuisance in old cities. In Holland, the "dear little" rabbit has become a rural pest which has to be shot, because they have multiplied so rapidly that they have damaged the dikes.

Attitudes to animals are illustrated by sayings about animals in general use: the horse is noble, the mule stubborn, the fox sly, the cat deceitful, the pig "stupid" (although this is not actually the case) and a person described as a monkey will be offended. The animal's use by humans defines the relationship between man and animal. A donkey may be greatly overburdened, since loading limits only apply to lorries. Wild rats have to be destroyed, while specially bred rats for animal research on the other hand, live in air-conditioned rooms, free of infection and fed on the best food. Man has always used the animal for food and clothing by yoking it to the plough or riding it to war. The direct relationship with the useful animal changed with industrialisation. The horse was replaced by the tractor. The neighbours' hens no longer bother us since they are now raised on a farm. Pets provide the best company to many. Statistics show that in and around Vienna, 100,000 dogs and 70,000 cats are fed with 21,000 tons of meat each year. Tinned food for dogs and cats does not consist of scraps, but according to modern scientific knowledge, is actually of high quality; in Austria almost a billion schillings is spent on them every year.

Specially bred animals are used to study the causes of disease or toxicity, since no useful scientific data can be obtained from inadequately cared for animals. This also results in useful new knowledge about the feeding and care of household pets and livestock, the study of their diseases and their treatment. In the 1900's, it was animal researchers who actually began working on the essential basis for the maintenance and care of animals.

With very active drugs, the protection of man from their possible adverse effects and toxicity is increasingly important. Thus, the licensing authorities set higher requirements on toxicity testing, which in turn require more and more animals. However, in 1976 a group of prominent European pharmacologists in a public statement "Towards a more rational regulation of the development of new medicines" (Conference in Sestri Levante, *Europ. J. Pharmacol* 11, 1977) proposed a revision of testing procedures and a limiting of toxicity investigations. This opened the way for the development of more specific and selective testing procedures. The breakthrough to new animal methods for toxicity testing, whether these are viewed as substitutes or alternatives, thus occurred rather earlier than is generally known.

If one leafs through years of the *Journal of Pharmacological Methods* one finds a continuing trend towards alternative methods. Furthermore, a new journal, *Toxicology in Vitro*, was inaugurated.

Analysis of the current situation
Thirty experts in medical research, from several European cities, discussed alternatives to animal research. All spoke in favour of the use of alternative methods, as long as their scientific value is demonstrable. In almost all areas of medical research a continuing move towards alternative methods has been taking place for years without external directives. Many of these methods have been routine for a long time. The predominant research animal is now the rat rather than the dog. However, the complete discontinuation of research on monkeys, dogs or cats is not possible. Even animal experiments without anaesthetic remain irreplaceable in certain narrowly restricted areas of research, although studies of this kind have long been performed under conditions which are not associated with suffering or pain for the animal.

The research scientist sees himself as a responsible member of society. The aim of his work lies in an improvement of safety and the quality of life of his fellow men. Animal researchers certainly do not see animal protectionists as their "opponents", especially since they themselves have contributed greatly to current understanding of maintenance, feeding and handling of animals. Only the unintelligent opponents of animal research are seen as adversaries, since they not only obstruct the dissemination of factual information, but also seek to incite credulous fellow citizens with unqualified opinions. Therefore, many of the contributions to this book have highlighted those areas of research which can only be performed with animals. Thus, anyone who understands the content and aim of a research project will also grasp the need for animal research.

Many people now see the potential hazards more than the actual advances in modern research - irrespective of the particular area. It seems to be similar with the motor car: when we did not have them we dreamt of them; if we do have them we talk about their costs and dangers. What biomedical research has brought to all of us is now a part of our lives and safety, both in general and in medicine. Even doctors undervalue the great benefits which have been granted us through animal experimentation. Often only few experts see the need for further developments. Their research aims, and the methods used to achieve them, are often difficult to explain, especially since researchers generally lack the ability to present the point of their work in a readily comprehensible way. Research personnel are designers and inventors, but not salesmen. As researchers, they are receptive to new basic ideas which are certainly valid for alternative methods as long as their value can be demonstrated.

The demand for alternative methods
Proof of scientific worth and reliability are of primary significance in the demand for alternative methods. This question has been discussed in almost every contribution. However, the *principle* of alternative methods is not solely the replacement of animal experiments. This would be as pointless as medical research taking place exclusively on volunteers or patients simply to avoid animal experiments. There are close mutual relationships between clinical research on patients, animal experiments and their alternatives. Alternatives to diagnostic or therapeutic measures also exist in clinical medicine. For example, ultrasound has often replaced X-rays. Modern drug treatment of ulcers has obviated gastric surgery in many cases. Chemotherapy replaced the difficult surgical treatment of tuberculosis used previously; prophylactic vaccination against tuberculosis gave immunity against the disease and thus, in turn, replaced chemotherapy.

Advances in clinical and experimental medicine run in parallel. Toxicology no longer simply describes the symptoms of poisoning, but represents the study of all aspects of poisoning and safety measures. Without animal experimental work this would be impossible.

Those who are young and healthy tend to be credulous, imagine that all medical problems will be solved and regard any further development of drugs as a luxury. However, then AIDS emerged, an infectious disease which is rapidly sweeping the world. Should society condemn this disease or is it a doctor's duty to study the condition? This is impossible, however, without research in monkeys. Without this animal research there would be neither treatment for the sick nor prophylaxis for the healthy!

The doctor's attitude to research is based on principles which sometimes differ from the general view of the individual: each of us has the right to decide whether he pursues sport or tooth care, whether or not he is inoculated and whether he eats more or less healthily. However, the doctor must decide on *ethical* grounds, since he is not only guilty if he commits wrong, but also if he omits to do the right thing by all his patients. This not only affects his direct dealings with patients, but also his attitude to medical research, including animal experiments.

Promotion of alternative methods
Alternative methods are not only promoted by those who imagine that pet dogs and cats are used for "agonizing experiments", but to a similar extent by those scientists who actually perform animal experiments. The former may be idealistically motivated, the latter have quite rational inducements for it.

Research is a perpetual search for the new, although the value of new knowledge must be proven. This also applies to alternative methods. Unfortunately, alternatives cannot be deliberately sought since they mainly result from unforeseeable methodological developments. Who would have foreseen at the beginning of this century that pernicious anaemia, which was then invariably fatal, could ever be treated? Who could have suspected that vitamin B_{12}, lack of which causes this disease, would one day be cheaply obtained in unlimited quantities as a by-product of antibiotic manufacture? Only the gradual acquisition of new knowledge enabled this advance to be made.

Old aims - new methods
If alternative methods are to be promoted, their discovery must be facilitated. However, this cannot happen if researchers are surrounded by "red tape". Research is a game which always has new possibilities; an uncluttered "play room" for testing new methods has almost always led to an important breakthrough. However, this free space will be lost if every experiment has to be questioned. It should not be overlooked that the researcher using animal experiments became an expert in stages, that his expertise was moulded by biochemical and physical methods and that he learned to work with animals or their organs from experienced colleagues. Only if he is methodically sound does he produce new results. Why have people so little confidence in his work? Only because he uses animals for research? People trust an electronics expert though, who seeks to develop a new type of transistor. Earlier animal experimental methods will soon be replaced by better, i.e. alternative, methods. Modern systems of scientific communication enable new methods to be disseminated very quickly. Within a year they are known all round the world, usually more rapidly than diagnostic or therapeutic innovations. Alternative methods are not kept secret, but are published, as are all experimental results. How often does a scientist return from visiting another laboratory having seen a new method which he then wishes to adopt?

Who pays for the delay if he has to wait months for permission to adopt these improved methods? Is not an experienced surgeon trusted, without official permission, to gradually improve an operation he has often performed? The animal researcher invests a great deal of time and personal energy. He does it willingly since he knows that his results will be at the disposal of everyone else. Every country which conducts research with experimental animals pays a tribute to medical knowledge which serves the wellbeing of all mankind. Drugs against tropical diseases were not developed in the tropics, but in countries where there were scientists who possessed the capability and the will to carry out their research.

Authorization for every single research project involving experimental animals offers only an *apparent* protection against unnecessary animal experiments, since these measures also hinder the development of new alternative methods. This is similar to making traffic problems worse by introducing parking restrictions! Two progressive and far-sighted moves would be:

1. Granting permission for specific types of activities in animal experimental research establishments.
2. Providing certificates of competence for the animal experiment researchers.

There are examples of this from industry and commerce; a new factory can only go into business if certain official regulations are adhered to. Accordingly, an experimental animal institute could be licensed to work only on isolated organs of dead animals, only on small laboratory animals such as rats, mice and guinea-pigs, or, if the appropriate conditions are fulfilled, also on larger research animals like cats or pigs. This is easy to establish and to officially control in a satisfactory manner.

The researcher's training, documented by publications or qualifications, can satisfactorily define which type of animal experiments he is qualified to perform. Similarly, the qualification of an electrician to work with electricity can be established - without his having to apply for permission each time he is to perform an installation. This approved control system can be utilised as a practicable, easily financed model for performing animal experiments. Thus, official registration like that of the factory inspectorate would exercise an effective function. At the same time, this would give space to the researcher who is trying to develop new methods, and thus to the aim to which we all strive, namely the promotion of alternative methods for animal research.

Glossary

A. Bucsics

Adrenaline	A hormone of the adrenal medulla released by stress. Syn.: epinephrine.
Aggregation	Conglomeration of cells, e.g. red blood cells. Syn.: agglutination.
Alkaloids	Natural substances containing nitrogen (most with complicated chemical structure), which are produced by many plants. They have an alkaline nature, hence the name. Examples: morphine, atropine.
Alzheimer's	Mental degeneration not usually occurring in old age, but often between 40 and 50 years.
Amniocentesis	Puncture of the amniotic sac to obtain amniotic fluid for the purposes of prenatal diagnosis
Amnion	The inner smooth sac protecting the unborn child.
Anabolics	Substances which influence metabolic structure (protein synthesis, growth factors, muscle composition). In general, the term is used to refer to synthetic substances resembling the male sex hormones.
Anaemia	A condition in which the haemoglobin concentration or red blood cell number is reduced below normal. Pernicious anaemia results from a defective uptake of vitamin B_{12}. Before the cause was known this disease was regarded as incurable, i.e.'pernicious'.
Anaesthesia	Insensitivity to pain, temperature or tactile stimuli.
Analgesia	Elimination of sensitivity to pain.
Analgesics	Pain-killing drug.

Glossary

Anastomosis	Congenital or artificially induced connection between two hollow organs, e.g. blood vessels.
Androgen(s)	Male sex hormones
Angina	Sudden pain in the thoracic region resulting from inadequate oxygen supply to the heart.
Anginal	That type of chest pain associated with shortness of breath and fear of death.
Antagonist	Substance whose effect is directed against any other (e.g. endogenous) substance.
Antibody	Proteins which are produced in an immune (defensive) reaction of special cells to a specific antigen, e.g. viral protein.
Antigen	Substance which triggers an immune (defensive) reaction in the body and reacts specifically with the products (antibodies) thus stimulated.
Antihistamine	A drug which inhibits or lessens the effect of histamine released in the body by certain diseases (e.g. allergies).
Antihypotensive	Agent for raising blood pressure lowered by illness.
Antiserum	Serum containing large amounts of a particular antibody.
Assay	Analysis of contents, detection method.
Atherosclerosis	Fat deposition on blood vessel walls ultimately leading to occlusion of the vessel.
Autopsy	Examination of the dead body to establish the cause of death. Syn.: Post-mortem, necropsy.
Bioassay	Detection method with a biological measuring system, e.g. on an isolated organ or a whole animal.
Biopsy	Removal of tissue for histological investigation on a living organism (e.g. by needle aspiration or during surgery).
Blocker	Substances which inhibit the interaction of certain (e.g. endogenous) substances with their corresponding receptor. For example, the so-called beta-blockers hinder the action of endogenous or exogenous noradrenaline and adrenaline on the beta-receptors of the heart.
Carcinogenic	Substances or factors which can cause cancer.
Cardioplegic	Designation for drugs which cause artificial cardiac arrest in open heart surgery.
Catheter	Tube which is introduced into body cavities or blood vessels in order to empty their contents or introduce fluids.
Chemotherapy	Inhibition of infectious pathogens or tumour cells in the body by chemical agents.

Glossary

Clinical	Concerning the observation and treatment of patients (as opposed to theoretical or experimental considerations).
Clone	The genetically identical progeny of a single cell, gene or organism.
Colostrum	The first milk produced by the mammary glands during pregnancy and available in the first 3-5 days after parturition.
Competitive	Describes the inhibition of biological processes by antagonists. In competitive inhibition, the antagonist (e.g. beta-blocker or coumarin) 'competes' for a binding site with the biologically active compound (e.g. adrenaline or vitamin K) thereby preventing the appropriate metabolic process.
Congenital	Abnormality caused before birth; present at birth.
Contraction	Refers to a muscle.
Coronary	The coronary blood vessels which supply the heart.
Coumarin	Orally-administered compounds which interfere with blood coagulation by inhibiting the action of vitamin K.
Cytostatic	The inhibition of cell growth and multiplication, and a term used to describe drugs used in cancer treatment.
Cytotoxic	Injurious to cells.
Decerebration	Transection of the spinal cord to eliminate brain function. Decerebration is done under anaesthetic and is a painless death for experimental animals.
Dementia	Intellectual disintegration which is not congenital.
Derivation	Production of a derivative of a compound, for example to confer certain desired properties on the substance.
Derivative	Chemical compound produced from another compound.
Dermatology	Study of the skin and diseases of the skin.
Diabetes	Generally used to describe 'diabetes mellitus', a chronic condition associated with increased blood sugar levels due to inadequate insulin production by the islet cells of the pancreas.
Diagnosis	Determination of the nature and/or cause of a disease.
Differential count	The staining of blood cells to differentiate the different types of white blood cells.
Digitalis	Drugs obtained from the Foxglove plant species, which strengthen the heart muscle.
Diuretics	Drugs which promote the excretion of salt and water from the body. They are also used for lowering high blood pressure.
DNA	Deoxyribonucleic acid: the chemical carrier of genetic information.
Dysfunction	A general term used to describe abnormal function.

Ectoparasites	Animal parasites living on humans or animals (e.g. fleas).
Electrolyte	Acids, bases or salts which decompose in aqueous solution to ions (charged particles). In medicine, the term usually refers to plasma levels of sodium, potassium and chloride ions.
Embryo	The phase of development from conception to (in humans) the end of the 8th week after fertilisation.
Embryotoxic	Any kind of toxic effect on the embryo or fetus (fetotoxic).
Endocrinology	Study of the function of hormones and the hormone-producing glands.
Endogenous	Describes a naturally-occurring substance. Endogenous depression has no external cause (e.g. misfortune), but is probably due to a chemical disturbance in the brain.
Endothelium	The smooth layer of cells coating blood vessels and serous body cavities (gastric and thoracic cavities, pericardium).
Enteritis	Inflammation of the bowel.
Enzyme	Proteins which accelerate chemical reactions in living organisms by functioning as catalysts.
Enzymatic	Describes chemical reactions which take place with the help of enzymes.
Epidemiology	The investigation of distribution of diseases, physiological variables and social incidence of disease in human populations as well as of factors which influence this distribution (WHO definition).
Erythrocytes	Red blood cells.
Eukaryote	Organism which has its genetic material contained in a nucleus separated from the rest of the cell by a nuclear membrane.
Exogenous	A substance which does not occur naturally in the body. (Opposite to endogenous)
Exudate	Efflux of fluids from blood or lymph vessels caused by inflammation.
Fetus	The unborn from the end of the embryonic phase to birth.
Fibrils	Thread-like structures forming part of a fibre.
Fibrin	The insoluble protein formed during blood coagulation.
Field test	Investigation which is carried out not under clinical or laboratory conditions, but on persons in normal surroundings and under relevant circumstances. Syn.: field study, field trial.
Fluorescence	The capacity of certain substances to emit light after irradiation (see also luminescence). The wavelength of the light emitted is typical for a given substance. Fluorescent

	substances can be chemically linked to other substances (e.g. antibodies) and render these detectable.
Gametocyte	Plasmodial (malaria-causing) cells, which produce male or female sex cells in the host (mosquitoes).
Gas chromatography	Process for separating and detecting gases and vaporisable fluids (whose boiling point is less than 350°C).
Gene	Functional unit of inheritance and genetic information.
Genetic	Concerned with heredity.
Genome	All the genes of an individual. Since each gene (apart from those in sex cells) is duplicated, the haploid (half) set of chromosomes and the genes localised therein is known as the genome.
Glycoside	Chemical compounds which contain sugar in a specific chemical concatenation. Digitalis is often referred to as a cardiac glycoside.
Gonadotropine	Pituitary gland hormone which regulates the activity of the reproductive glands.
Haemodynamics	Describes the physical factors involved in blood flow within the circulation.
Histamine	Endogenous substance which is produced by mast cells, and released during allergic reactions. It is responsible for many of the typical symptoms of allergy.
Histology	Study of the body's tissues.
Homeopathy	Theory expounded by S. Hahnemann for the treatment of diseases, which prescribes low doses of substances which, in higher doses, provoke symptoms similar to the disease in healthy people.
Homogenate	Very finely reduced tissue usually in aqueous suspension. Cell components are released by the process of homogenisation.
Hybridisation	A term used in molecular genetics to indicate the formation of double-stranded nucleic acid molecules from single-stranded polynucleotides. The technique may be used to introduce foreign genes into the host DNA.
Hypertension	High blood pressure.
Hypoglycaemia	Low levels of blood sugar.
Hypophysis	The pituitary gland, a cherry-sized structure at the base of the skull which consists of two lobes in humans. The anterior lobe produces growth hormone, regulates the activity of different glands (thyroid, adrenal, sex, mammary) and is itself regulated by the cerebral region of the hypothalamus. The posterior lobe

	accumulates and releases the hormones produced (vasopressin, oxytocin) into the hypothalamus as needed.
Hypophysectomy	Removal of the hypophysis.
Hypothalamus	Part of the brain where the coordinated regulation of the most important regulatory processes takes place e.g. temperature, blood pressure, breathing.
Hypotension	Low blood pressure.
Immunise	To produce antibodies against an infectious disease in an organism by introducing the appropriate antigen.
Immunobiology	The branch of biology concerned with immunological phenomena.
Immunology	The study of an organism's detection and defence mechanisms for exogenous and, under certain conditions, endogenous substances.
Immunosuppression	The reduction or suppression of the immune response.
Infection	Transmission, adhesion, penetration and growth of microorganisms (e.g. fungi, bacteria, viruses) into a host (plants, animals, man). This is necessary for the occurrence of an infectious disease, although disease does not necessarily occur after every infection.
Infusion	Input of large amounts of fluid, e.g. of blood or drugs diluted with glucose or saline solutions, usually into a vein. Infusion is generally performed dropwise.
Inoculate	Introduction of infectious material, e.g. into the body for immunisation, or into nutrient medium for culturing.
In situ	In its natural location (e.g. the position of an organ within the body).
Intact animal	Whole animal experiments and trials on intact animals. Whole animal animals defines investigation on living animals. See also *in vivo*.
Invasive	Methods of investigation or treatment associated with surgical intervention, or a term used to describe tumour growth into neighbouring tissue.
In vitro	'In the test tube', i.e. performed outside the organism. Opposite to *in vivo*.
In vivo	Experiments performed on a living organism. Opposite to *in vitro*.
Leukaemia	Disease due to disordered or increased production of white blood cells.

Luminescence	Emission of light by certain chemicals after irradiation with electromagnetic (light) waves or particles. With fluorescence, the light emission occurs immediately after irradiation, while light emission over a longer space of time after irradiation is characterised as phosphorescence.
Lymphosarcoma	Malignant lymph node enlargement.
Macrophages	Cells which can move like amoebae, which also circulate in the blood (monocytes) and take part in the body's defence mechanism.
Malignant	The malignant course of a disease. Opposite to benign.
Mast cells	Cells present in blood (basophils) and in tissue (tissue mast cells). Their granules contain histamine, serotonin and heparin-containing granules, which may be released by certain antigen-antibody reactions.
Matrix	Parent tissue, basic substance, in which moulding and other typical elements are deposited.
Merozoites	Plasmodial forms, which can multiply asexually in the red blood cells of malaria victims.
Mesenterial	Blood vessels of the intestinal mesentery.
Metabolism	The breakdown, often enzymatic, of drugs and other substances within the body into more easily excreted products (metabolites). These can be either less toxic (detoxification) or more toxic (toxification).
Microbiology	Study of microorganisms.
Microfilariae	First larval stage of threadworms, which are transmitted into humans in the tropics by insects.
Mitral	Valve between the left vestibule and the left valve ventricle.
Molecular biology	Study of the molecular bases of life. Modern molecular biology is widely concerned with the molecular bases of heredity (molecular genetics).
Monoclonal	An antibody derived from a single clone of immunoglobulin-producing cells.
Morphine	An important active agent in the poppy with pain-killing and sleep-inducing properties. Can lead to addiction especially if processed into heroin.
Morphology	Study of the structure of the body and its organs and forms.
Myeloma	A malignant tumour of the bone marrow associated with plasma cell proliferation and excessive immunoglobulin production.
Myocardial	Refers to the heart muscle.

Neoplastic	Newly formed; in the narrower sense indicates tumour formation.
Nerves, peripheral	This refers to the nerve filaments which supply the individual organs of the body, especially as opposed to the central nervous system.
Neuralgia	Pain in the area innervated by one or more nerves.
Neurobiology	Biology of the central and peripheral nervous system.
Neuroleptics	Sedative drugs which are used in the treatment of mental disorders. Syn.: antipsychotics.
Neurotransmitter	Substances which, as opposed to electric stimulation within the nerve cells, conduct the impulse from one nerve ending to other or to the receiving organ (e.g. muscle cells) by chemical means. Examples: acetylcholine, noradrenaline.
Noradrenaline	Transmitter substance of the sympathetic nervous system and of certain brain cells. It is stored in vesicles in the nerve endings and released by excitation of the nerves. Noradrenaline together with adrenaline is present in the adrenal medulla.
Nucleotide	The pentose sugar and phosphate molecule in combination with either a purine or pyrimidine base, which in long chains, form nucleic acids (ribonucleic acid and deoxyribonucleic acid), the building blocks of genetic information.
Nucleus	The membrane bound structure at the cell centre containing the chromosomes and directing the metabolic processes of the cell.
Oedema	Painless, non-reddened swelling as a result of accumulation of fluid in the tissue. Oedemas can be localised (circumscribed) or generalised (affecting the whole body), Syn.: dropsy.
Oestrogens	Steroid hormones produced in the ovaries, placenta and in smaller amounts in the testicles. They help to regulate reproduction. Syn.: follicular hormones.
Oncofetal	Designation for compounds (antigens), which can be detected during fetal development as well as in tumour cells.
Organelles	Structures within the cells which are delineated by membranes and which usually have a special function (e.g. microchondria, chloroplasts).
Organogenesis	Phase of embryonic development in which essential organs are formed, and in which teratogenic effects can be triggered.
Osteoporosis	Quantitative reduction of bone tissue in existing bone structure.

Glossary

Papilliary muscle	Cone-shaped muscle protrusions on the inner walls of the heart. They are connected to the heart valves by thread-like tendons.
Parenteral	By-passing the gastro-intestinal tract, e.g. by injection.
Parkinson's disease	Lack of certain (dopamine-containing) cells in the brain. The most significant symptoms are hypokinesia, muscle rigidity and tremor.
Pathogen	An agent causing disease.
Pathogenesis	The process or cause of disease.
Pathology	Study of diseases (causes, occurrence, course as well as the physical symptoms and changes).
Peptides	Compounds which, like proteins, consist of amino acids, but are shorter. In the body they fulfil numerous messenger functions (hormones, neurotransmitters).
Perfuse	To circulate through the body or a part of the body, e.g. with a nutrient fluid.
Peristalsis	Undulating movement of the stomach, intestines, vas deferens and ureter to move the contents on.
Per os	To take by mouth. Opposite: parenteral.
Pharmaceuticals	Drugs
Pharmacokinetics	Study of the behaviour of drugs in the organism including their uptake, distribution in the body, metabolism and excretion.
Pharmacology	Study of drugs.
Physiology	Study of normal bodily processes.
Plasmodia	Malaria pathogens: Family of sporozoa which primarily attacks red blood cells.
Polyclonal	Stemming from many cells. Usually refers to an antibody produced by conventional immunisation techniques. Opposite to monoclonal.
Potency	Activity of a defined dose of drug or chemical compound.
Prenatal	Before birth.
Progestagens	Hormones which are produced after ovulation in the ovary and placenta. They regulate, together with oestrogens, the monthly cycle and help maintain pregnancy. Syn.: progresterone.
Prokaryote	Organism without a nucleus in which the genetic material is not separated by a nuclear membrane from the rest of the cell, e.g. bacteria. Opposite: eukaryote.
Prostaglandins	Collective term for hormone-like substances with many different biological effects (e.g. on inflammation, blood

	coagulation, inducing labour etc.). They are produced in all organs from arachadonic acid.
Proto-oncogenes	Precursors of oncogenes, i.e. factors which (can) cause tumour formation through the effect of other factors.
Psychotropic	Affecting mental processes (e.g. drugs or alcohol).
Radioassay	Detection and assay methods which use radioactively labelled chemical substances.
Radioimmunoassay	*In vitro* assay methods based on an antigen-antibody reaction with a radioactively labelled antigen.
Receptor	An are of a molecule or cell which can bind specifically to another molecule or cell.
Schizonts	Life stage of malaria parasites which mostly disintegrate into merozoites.
Schizophrenia	Split personality; an endogenous mental illness of unknown origin.
Screening	A term usually applied to the progressive (pre-) investigation of a series of substances or group of persons for desirable or undesirable effects, and the application of tests for detecting early stages of disease.
Secretion	Discharge. This can take place to the outside, as for example with glands with efferent ducts, or into the circulation as with hormone-producing glands.
Sensitisation	Antibody formation after antigen contact.
Sequencing	Determination of the order of amino acids making up a protein polypeptide chain or DNA.
Serotonin	A mediator and transmitter substance present in intestinal mucous membrane, blood platelets, parts of the brain and certain gastric tumours. The determination of serotonin metabolites in urine may be utilised for the detection of these tumours. Syn.: 5-hydroxytryptamine.
Shock	Failure of peripheral circulation with lack of blood flow in the smallest blood vessels. Anaphylactic shock is caused by an antigen-antibody reaction.
Sclerosis	Pathological hardening of an organ.
Spasmolytic	Drug relieving cramps.
Sporozoite	Infection stage of sporozoa, e.g. of malaria or toxoplasmosis pathogens.
Substitution therapy	Treatment of a disease by replacing missing substances which are normally present in the body, e.g. defective thyroid gland function by giving thyroid hormone.

Suppressive therapy	Treatment of a disease by a measure which suppresses the disease symptoms but does not eliminate the basic cause.
Synthesise	Artificial production of a chemical compound from elements or more simple matter.
Scintigraphy	Imaging parts of the body localised with the gamma rays of externally administered radioactive compounds which accumulate in the target organ.
Teratogenic	Ability of an agent to cause congenital gross structural abnormalities (congenital anomalies or 'malformations').
Teratology	The study of malformations.
Therapeutic	Agent for treating an illness.
Thrombosis	Clotting of blood inside the blood vessels of a living organism.
Thrombocytes	Blood platelets.
Toxic	Poisonous.
Trachea	Wind pipe.
Traumatisation	Acquiring of injuries.
Tumourigenesis	The first step in tumour causation.
Validation	Conclusive proving of the usefulness of a new method.
Vasopressin	Hormone produced by the posterior pituitary lobe which contributes to water and salt regulation. The name comes from the sharp rise in blood pressure which large quantitites of the exogenous hormone are able to effect.
Vector	A term for disease-transmitting insects, and in the broader sense, for other means of disease communication.
Ventricular muscle	Muscle of the ventricle of the heart.
Ventricular septum	Dividing wall between the right and left ventricle.
Vivisection	'Cutting up living animals', i.e. without anaesthetic, as performed before the discovery of anaesthetics, but with muscle-paralysing agents. It is now forbidden on ethical grounds.

Index

Acetylsalicylic acid 87
ACTH 114, 115, 173
AIDS 124, 156, 163, 186
Allergy 10, 46, 47, 235
Alzheimer's disease 177, 178, 231
Anaesthetics 18-19, 25, 103, 199, 241
Analgesics 25, 71, 181, 183, 231
Anaphylactic shock 10, 240
Angina pectoris 32, 89, 232
Animal anaesthesia 190 (see also anaesthetics)
Animal maintenance 192-194, 206, 218, 220, 221-225
Animal protection law 137-138, 142, 212, 215-217
Antagonists 10, 117, 197, 203, 232, 233
Antibiotics 14, 25, 27-28, 128
Antibodies 76, 77, 105, 115, 135, 150, 154, 155, 156, 158, 159-160, 162, 187, 188, 200, 205, 232, 235, 236
Antigen 10, 105, 115, 116, 148-150, 155, 158, 186, 187, 232, 236, 237, 240
Antigen-antibody reaction 10, 115, 240
Antihistamine 10-11, 232
Antipsychotics 238
Aspirin 87
Autoimmune diseases 158
Avermectine 189-190

Beri-beri 13
Beta-blockers 22, 32, 33, 232
Beta-receptors 22, 32, 33, 232

Bioassay 71, 112-116, 232
Biotransformation 41, 44, 54, 59
Blood coagulation 11-12, 17, 87, 107, 202, 233, 234, 240
Blood pressure 7, 21-23, 42, 69, 71, 89, 118, 121, 195, 205, 223, 233, 235, 236, 241
Blood sugar 14, 19, 20, 40, 112-114, 195, 233, 235
Brain functions 174-179
Bromocryptine 25

Cadaver skin 103
Calomel 18
Carcinogenicity 18, 46, 51, 59, 128-131, 140, 141, 142, 144, 167,
Carrageen 183
Castration 11, 111
Cats 7, 8, 28, 33, 36, 62-64, 90, 184, 190, 206, 220-222, 225, 226, 228, 229
Cell culture 54-60, 78, 79, 81-83, 85, 104, 146, 149, 150, 158, 160, 162, 163, 177, 183, 200, 204
Cell-free models 41
Chemoprophylaxis 122, 128
Chemotherapy 15, 26, 40, 57, 121-123, 125, 133, 145-150, 197, 232
Chimpanzees 163, 171, 174
Cholesterol metabolism 162
Chromosome aberration 43
Cloning 151, 186
Computer 24, 48, 55, 66, 98-99, 106-108, 121, 123, 199
Contergan 20
Contraceptives 112, 113
Corticosteroids 112, 192
Cortisone 25, 110, 160, 161
Cosmetics 44-47, 54, 57, 65, 215
Curare 22, 65, 90, 195
Cytostatics 133, 134, 145-148, 150, 152

Databanks 98, 163
Depression 170, 177, 179, 234
Detoxification 43, 66, 104, 142, 144, 237
Diabetes 14, 19, 23, 42, 112, 157, 233
Diethylstilboestrol 130
Digitalis 19, 25, 233, 235
Diuretics 23, 26, 30, 37, 233
Dogs 7, 8, 11, 14, 28, 31, 33, 36, 62-64, 66, 67, 89, 112, 117, 138, 149, 169, 185, 190, 201, 202, 206, 218, 220-223, 225, 226, 228

Draize test 46-47, 57, 102-103
Drug development 14, 17, 29, 71, 72, 118, 177
Drug safety 38-44, 166, 219

ELISA 116
Embryo 15, 42, 51, 66, 82, 167, 234
Embryotoxic 54, 66, 164-165, 234
Endocrinology 109-116, 178, 234
Endoscopy 198, 200
Endothelial cells 75, 204
Enteritis 184, 186, 234
Epidemiology 53, 129, 130, 165, 234
Ethics 52, 82, 93, 94, 207, 212
Ethopharmacology 169, 172-174
Feline leukaemia 186
Foot and mouth disease 187
Frog heart 8, 70, 118

Genetic defects 5, 16, 76, 83, 133, 147
Genotoxicity 51, 52, 59, 131, 132
Guinea-pigs 7, 16, 28, 36, 46, 66, 70, 95, 112, 115, 159, 170, 202, 229

Heart 34, 70, 107, 118
Heart surgery 197, 201, 232
Heart-lung machine 202-203
Heparin 12, 202, 237
High pressure liquid chromatography (HPLC) 105
Homogenate 7, 235
Hormones 12, 14, 21, 25, 39, 64, 65, 73, 75, 77, 90, 105, 106, 110-116, 128, 158-162, 231, 232, 234, 236, 238, 239
Human chorionic gonadotrophin (HCG) 159-160
Hybridoma 77, 160

Immune cells 143, 154, 186
Immune system 57, 85, 147-149, 153-158, 160-162, 186
Immunisation 75, 186, 236, 239
Immunology 59, 142, 144, 153-163, 220, 236
Immunomodulation 157
Immunotoxicity 58
Industrial chemicals 47, 52, 54, 57, 100

Infection 6, 15-17, 28, 122-126, 128, 132, 149, 153-155, 161, 163, 197, 200, 205, 215, 225, 236, 240
Infectious diseases 57, 95, 121, 125, 153, 159, 186
Information systems 29
Inhalation toxicity 57
Insulin 14, 19, 25, 73, 74, 105, 113, 114, 157, 195, 196, 233
Interferon 85, 148, 157
Interleukin 157
Iodine deficiency 110

Joint inflammation 181, 183

Killing for research purposes 216
LD_{50} 11, 15, 46, 57, 100-103
Lipstick 44, 45
Lithium 170
Liver toxicity 46
Local anaesthetics 25, 103 (see also anaesthetics)
Lymphocytes 143, 148, 150, 154-156, 161-163, 189

Maintenance 6, 63, 64, 76, 77, 137, 140, 184, 189, 191-193, 206, 218-225 (see also animal maintenance)
Malaria 14, 89, 95, 122, 124-128, 235, 237, 239, 240
Material tests 199-200
Maximum Workplace Concentrations (MWC) 51-52
Membrane preparations 178, 183
Metabolism 34, 40, 42, 54, 55, 58, 60, 85, 123, 132, 142, 163, 188, 189, 201-203, 237, 239
Microsurgery 64, 200
Monkeys 36, 62, 63, 66, 127, 171, 196, 206, 222, 226, 227
Monoclonal 143, 146
Monoclonal antibodies 77, 127, 150, 159-160, 237, 239
Morphine 11, 71, 180, 191, 196, 231, 237
Muscle relaxants 22
Mutagenicity 46, 51, 56, 59, 144
Mutation 83, 134
Myocardial cells 81

Neurotoxins 176
Neuraminidase 149-150
Neurotransmitter 238

Nude mice 133, 134, 146, 147, 152
Nutrient media 70, 123

Onchocercosis 190
Oncogenes 134-135, 240
Operation, surgical 9, 77, 140, 182, 197, 198, 201, 203
Organ cultures 56, 85

Pain 5, 10, 11, 19, 23, 24, 29, 32, 46, 65, 71, 72, 82, 89, 103, 105, 109, 138, 140-143, 174, 179-184, 192, 197, 198, 208, 209, 212-215, 231, 232, 237, 238
Parasite control 189-190
Parkinson's disease 9-10, 177-178, 239
Penicillin 15, 23, 25-27, 29, 42, 90, 91, 190, 198
Personal licence 216
Pests 62, 140, 141, 212, 220, 221
Pharmacodynamic 44
Pharmacokinetic 44, 53
Pigs 7, 16, 28, 36, 46, 62, 65, 66, 70, 72, 74, 78, 95, 112, 115, 119, 158, 170, 190-194, 202, 221, 222, 229
Plastic surgery 197
Postmortem 75, 77, 78
Pregnancy test 160
Prenatal diagnosis 76, 231
Project licence 216, 217
Prontosil 16
Prostaglandins 87, 160, 189, 239
Prothrombin 11-12
Pure research 63-65, 92, 105, 184, 188, 221

Quick test 11-12
Quinine 15, 89, 124

Rabbits 7, 11, 14, 15, 19, 33, 36, 46, 47, 64-66, 103, 106, 111, 115, 119, 160, 192, 195, 223
Radioimmunoscintigraphy 200
Rat poisons 12
Receptors 22, 28, 32, 33, 76, 135, 177, 232
Reproduction toxicity 58

Safety tests 66
Schizophrenia 177-179, 211, 240

Screening 26-29, 56, 57, 59, 91, 152, 153, 173, 177, 240
Simulation 64, 65, 106-108
Slaughter house material 73
Species-specific 139, 185, 189
Standards 41, 56, 113, 166, 184
Stereotypes 191
Stress 23, 54, 106, 110, 117, 121, 173, 192-194, 231
Substitution therapy 14, 240
Supplementary methods 69, 158

Taboos 82
Teratogen 67
Therapeutic index 15
Thrombosis prophylaxis 12
Tissue cultures 60, 68, 84, 85, 126
Toxicity 15, 26, 41-44, 46, 50, 51, 54, 56-58, 60, 64, 67, 95, 100, 102, 104, 124, 129, 132, 144, 145, 168, 208
Toxicity testing 57, 64, 102, 124, 208, 226
Toxicology 38, 39, 46, 48, 50, 52, 54-58, 60, 66, 83, 84, 98, 100, 102, 104, 132, 142, 164, 166-168
Toxification 43, 143, 237
Tranquilliser 91, 190
Transformation 43, 52, 56, 59, 74, 84, 123, 130, 131, 134, 142, 146
Transplantation 7, 76-78, 147, 157, 199, 201, 204
Tuberculosis 27, 155, 188, 194, 227
Tumour cells 57-59, 85, 133, 134, 141, 143, 146, 148-150, 156, 160, 200, 232, 238
Tumour immunology 143
Tumour localisation 77
Tumour necrosis factor 148
Tumour therapy 77, 134, 144, 145-150
Tumour-promoting effects 131

Vaccines 85, 145, 159, 184, 186-188
Vasopressin 16, 192, 236, 241
Veterinary medicine 184, 185, 189-193
Vitamin research 13-14
Vivisection 17, 92, 97, 195, 197, 207, 210, 241